AF801039

Contributions to Finance and Accounting

The book series 'Contributions to Finance and Accounting' features the latest research from research areas like financial management, investment, capital markets, financial institutions, FinTech and financial innovation, accounting methods and standards, reporting, and corporate governance, among others. Books published in this series are primarily monographs and edited volumes that present new research results, both theoretical and empirical, on a clearly defined topic. All books are published in print and digital formats and disseminated globally.

More information about this series at https://link.springer.com/bookseries/16616

Hasan Dinçer • Serhat Yüksel
Editors

Sustainability in Energy Business and Finance

Approaches and Developments in the Energy Market

Springer

Editors
Hasan Dinçer
Faculty of Economics and Administrative Sciences
Istanbul Medipol University
Istanbul, Turkey

Serhat Yüksel
Faculty of Economics and Administrative Sciences
Istanbul Medipol University
Istanbul, Turkey

ISSN 2730-6038 ISSN 2730-6046 (electronic)
Contributions to Finance and Accounting
ISBN 978-3-030-94050-8 ISBN 978-3-030-94051-5 (eBook)
https://doi.org/10.1007/978-3-030-94051-5

© The Editor(s) (if applicable) and The Author(s), under exclusive license to Springer Nature Switzerland AG 2022
This work is subject to copyright. All rights are solely and exclusively licensed by the Publisher, whether the whole or part of the material is concerned, specifically the rights of translation, reprinting, reuse of illustrations, recitation, broadcasting, reproduction on microfilms or in any other physical way, and transmission or information storage and retrieval, electronic adaptation, computer software, or by similar or dissimilar methodology now known or hereafter developed.
The use of general descriptive names, registered names, trademarks, service marks, etc. in this publication does not imply, even in the absence of a specific statement, that such names are exempt from the relevant protective laws and regulations and therefore free for general use.
The publisher, the authors and the editors are safe to assume that the advice and information in this book are believed to be true and accurate at the date of publication. Neither the publisher nor the authors or the editors give a warranty, expressed or implied, with respect to the material contained herein or for any errors or omissions that may have been made. The publisher remains neutral with regard to jurisdictional claims in published maps and institutional affiliations.

This Springer imprint is published by the registered company Springer Nature Switzerland AG
The registered company address is: Gewerbestrasse 11, 6330 Cham, Switzerland

Contents

Creation of Energy Risk Insurance System

Laura Baitenova, Lyailya Mutaliyeva, Natalia Sokolinskaya, and András Vincze

1 Introduction

The importance of creating a system that protects people's energy risks in banking institutions was recognized at the end of the last century. In the Decree of the head of state, the Bank of the Russian Federation was instructed to speed up the formation and launch of the fund for insuring the financial assets of Russian banks to protect project funds (Qiu et al., 2020). Now, energy risk insurance (or the energy risk insurance system) is an important, relevant, and mandatory system, due to the efficiency of which the stability of the economy in the state is ensured (Zhou et al., 2021). The most vulnerable and fragile, especially in times of crisis, is the work of economic market agents, the efficiency of monetary and credit institutions, as well as other intermediaries whose functions are significant for the formation and improvement of the state of the economy and objects of economic activity (Fang et al., 2021).

The most urgent problems are those related to providing financial protection to creditors, especially small companies and individuals, whose behavior can lead to

L. Baitenova (✉)
Almaty University of Power Engineering and Telecommunications named after Gumarbek Daukeev, Almaty, Kazakhstan

L. Mutaliyeva
L.N. Gumilyov Eurasian National University, Astana, Kazakhstan

N. Sokolinskaya
Financial University under the Government of the Russian Federation, Moscow, Russian Federation

A. Vincze
Circular Economy University Center, Renewable Energy Research Group, University of Pannonia, Veszprém, Hungary

© The Author(s), under exclusive license to Springer Nature Switzerland AG 2022
H. Dinçer, S. Yüksel (eds.), *Sustainability in Energy Business and Finance*, Contributions to Finance and Accounting,
https://doi.org/10.1007/978-3-030-94051-5_1

confusion, destruction of the fragile balance of the entire system, and bankruptcy. An important condition for the operation of any banking system is the presence of trust in it on the part of projects. Just insurance of energy risks in various banking organizations and in the state inspires projects with confidence that in case of any unforeseen situations, they will be able to return either the entire amount of funds invested in the bank, or at least part of the amount, but not lose everything that was given to the bank for storage and accumulation.

Of course, payments to projects in the event of bank bankruptcy are made at the expense of special funds created by banks with the active participation of the government and the state. For the most part, energy risk insurance is aimed at protecting small projects that do not have any opportunities and means to own information about the bank. Those projects that do not have sufficient information about the bank, deposits, and various financial processes are usually unable to study, monitor, or analyze the offers of competing banks to assess their level of reliability due to a lack of incoming information.

It is worth considering that the obligation to insure energy risks appeared on the territory of Russia relatively recently. This point is explained by the fact that in Soviet times there was no need for such an organization, all banks were officially considered state-owned banks at that time, and the state itself, in turn, gave guarantees that the funds invested by citizens were safe. The energy risk insurance, working not only in Russia, but also in other countries, was able to prove its usefulness, efficiency, and sufficient level of quality, which was also proved by the assistance of the system and its representatives to the fact that various elements of credit resources, that is, citizens 'money, are involved in this section.

Everything listed above also explains not only the importance, but also the relevance of the research work. The object of research is the peculiarities of formation of energy risk insurance. The subject of the research is both economic and financial relations, as well as natural rules of operation of energy risk insurance of citizens with non-profit banking organizations in Russia. The purpose of the study is to identify both positive and negative aspects of the operation of the energy risk insurance system, as well as to determine the prospects for the operation of energy risk insurance in the future.

Methodological basis of the study. During the research and its formation, private scientific and general scientific methods and sources were used. Also, when creating the work, we used an analysis of the works of researchers and scientists in the relevant fields. One of the main (theoretical) implications of this study is its results, which expand and enrich the understanding of the formation of insurance protection of financial energy risks of citizens.

2 Literature Review

As for the practical significance of this work, it is an opportunity to use recommendations, suggestions, and conclusions in the subsequent work of banking institutions to improve the overall performance of insurance coverage. In any modern state, there is a well-developed mechanism that protects money (Conteh et al., 2021; Denisova et al., 2019; Huang et al., 2021a, 2021b; Mikhaylov, 2018a, 2018b, 2022; Mikhaylov et al., 2019; Meynkhard, 2019, 2020; Nyangarika et al., 2019a, 2019b).

The basis of this system is, for the most part, that even if the bank goes bankrupt (or its license is revoked), the obligation to return money to projects will be transferred to a special organization. The main task of the system is to protect the personal and financial interests of even the smallest projects (Alwaelya et al., 2021; An & Mikhaylov, 2020, 2021; An et al., 2019a, 2019b, 2020a, 2020b, 2020c, Dooyum et al., 2020; Grilli et al., 2021; Gura et al., 2020; Mikhaylov, 2020a, 2020b, 2020c, 2021a; Mikhaylov & Tarakanov, 2020; Mikhaylov et al., 2021a, 2021b; Moiseev et al., 2020, 2021; Morkovkin et al., 2020a, 2020b; Mutalimov et al., 2021; Varyash et al., 2020; Yumashev & Mikhaylov, 2020; Yumashev et al., 2020; Zhao et al., 2021).

From the very beginning of its existence, insurance has been one of the most important methods of ensuring the need to compensate for damage in the event of unforeseen events (An et al., 2021; Danish et al., 2020, 2021; Dayong et al., 2020; Ivanyuk, 2018; Ivanyuk & Berzin, 2020; Ivanyuk & Levchenko, 2020; Ivanyuk & Soloviev, 2019; Ivanyuk et al., 2020; Lisin, 2020; Mikhaylov et al., 2018; Nyangarika et al., 2018, Uyeh et al., 2021).

Insurance as a process can be viewed from several points of view:

Economic: In this case, insurance is an economic relationship formed during the creation, distribution, and use of public trust funds necessary to compensate for losses, if they were received as part of an insured event. Refunds are made on a contractual basis (Bhuiyan et al., 2021; Candila et al., 2021; Dong et al., 2021; Dorofeev, 2020; Liu et al., 2021a, 2021b, 2022; Mikhaylov, 2021b, Mukhametov et al., 2021; Radosteva et al., 2018; Ranjbar et al., 2017; Rathnayaka et al., 2018; Saqib et al., 2021; Sunchalin et al., 2019; Udalov, 2021; Uandykova et al., 2020; Yüksel et al., 2021a, 2021b, 2021c).

Financial: In this case, insurance becomes an autonomous financial institution, which represents a whole complex of economic relationships, under which financial insurance funds are created (Mikhaylov, 2018c; Mikhaylov et al., 2019). In order to handle different financial risks, this situation becomes a necessity (Jun et al., 2021; Kou et al., 2021; Silahtaroğlu et al., 2021).

Legal information: Insurance here is a set of legal norms through which social relations are regulated, which are manifested during both the creation and use of insurance funds. Insurance can also be studied as a contract, a specific legal obligation, and a legal relationship (Melnichuk et al., 2020; Nie et al., 2020). Insurance is a certain type of legal relationship in which the insured person pays the insurance company a certain amount of money. In exchange for this, in the event of an insured event, the insured person will receive compensation for possible financial losses from the insurer.

In such a relationship, the party–the insurer will have to bear the risk for some time for the consequences that negatively affect the property or life of the insured person (policyholder) in the event of insured events. Upon the occurrence of events, the insurer must pay the other party the insurance amount.

The energy risk insurance adheres to several fundamental principles in its education:

- Transparency in the implementation of activities (Li et al., 2020).
- Accumulative nature, which is achieved due to the constancy of contributions (Yuan et al., 2021).
- Mandatory participation (Liu et al., 2021a, 2021b).
- Reduction of the level and magnitude of risks for projects if banks did not fulfill their obligations in emergency cases.

3 Methods

If we adhere to the concept of formation of energy risk insurance, then certain relationships in energy risk insurance appear based on legal norms, and not because of the free expression of the will of the parties. Also, these relations develop based on legislation and end their existence since the same legislation. The relationships created in energy risk insurance are based on two principles—subordination and power, which means that the relationships related to the formation, distribution, and use of the energy risk insurance fund are property-based.

Today, insurance can be considered as both a social guarantee provided by the state and a source for investment. One of the key tasks assigned to energy risk insurance is to protect citizens 'funds placed by citizens themselves in banks. In many countries, there is a system for protecting the financial condition and interests of the population, which is perhaps the most important social task. Energy risk insurance is mandatory in any member State of the European Union. As an example, energy risk insurance operates on the territory of Brazil, the USA, Japan, as well as on the territory of the CIS countries-Armenia, Ukraine, Kazakhstan, and others.

In general, it is possible to classify existing energy risk insurances in the world according to numerous criteria. A method for organizing energy risk insurance. In this classification, there are systems with positive guarantees, as well as systems with those guarantees that are not clearly expressed. The essence of such systems is revealed in the following names:

- Legal guarantee (usually they are also called insurance systems).
- General state guarantees (sometimes they are also called guarantee guarantees).

A characteristic feature of the first type is the existence of a procedure established at the legislative level concerning compensation for possible financial losses to projects in the event of bankruptcy of a banking institution that is part of energy risk insurance. Clients of a banking institution will know in advance about the availability and amount of the insurance amount available to them in case of problems in the bank's operation. Such a system allows you to create confidence in projects that their money will be saved at the expense of predictability. Also, such a system allows you to accumulate free financial resources of a banking organization.

A characteristic feature of the second type of system is the lack of strict legislative regulation, which determines the methods and possible ways to protect energy risks. The possibilities for obtaining compensation, as well as the amount of this compensation itself, depend entirely on the current situation, as well as on the decision made by the state bodies that determine the terms, conditions, and amounts of payments. The basis of such a system is trust in the state on the part of citizens, which is also a characteristic feature of states that have centralized strict management and differs in the dependence of banking institutions on various structures of state activity.

Different countries have different ways of addressing issues related to the use of existing financial investment guarantee systems. For example, Australia and New Zealand do not have any insurance systems at all, but instead of insuring energy risks, these countries have established disclosure requirements, which strengthen economic controls. Organization of participation of banking institutions in insurance systems. Within the framework of this classification, it is possible to distinguish systems of mandatory and voluntary participation of banking institutions.

4 Results

Usually, "money transfer operations" are carried out in non-cash form, unless, of course, the bank has signs of insolvency. On the territory of Russia, the creation of energy risk insurance is associated with the need to:

- Solving the constitutional priority tasks of the state related to the protection of citizens 'rights and guaranteeing their peace of mind.
- Creating prerequisites for increasing the overall level of people's trust in banks.

Today, there are no systems of banking institutions that do not have the risk of facing a crisis, just as there is no energy risk insurance that could be suitable for all banks in the entire state. Both the formation and development of energy risk insurance in Russia took place in several different stages. The creation of energy risk insurance, as well as the development of the regulatory and legislative framework, took 10 years, and the process itself was quite difficult.

It is noteworthy that experts from the USA, England, and Switzerland were involved in creating such laws, as well as in forming proposals related to the protection of individual savings. For this reason, the domestic legislation of those years was based mostly on European and American practice. Due to the fact that banks in those years were just beginning to be created, and no one not only knew about any bankruptcy, but also did not think about it, bank managers did not openly support this idea. But, even with this in mind, funds were still collected, although no one used them.

Ideologists who were at the origin of the formation of energy risk insurance recalled that this idea, as well as its promotion, was extremely difficult to move forward. But even at the beginning of the foundation's formation, everyone realized the importance of forming such a system, although at the initial stages, there were disagreements on some issues, especially regarding the financial content of the fund. Given the persistent and noticeable budget deficit, the high level of inflation, and the importance of correcting the economic situation in the country, issues related to restoring people's confidence in banks and in the banking system turned out to be important.

This very decree laid the foundation for energy risk insurance for the first time in Russia's history. The same decree also determined sufficient protection of the interests of physical projects, also indicating the protection of citizens 'savings with measures aimed at creating energy risk insurance as an analog of systems existing in other states. However, this work was never carried out. According to experts, this step, although it was taken, still violated the legislation in force at that time, as well as the charter of the Central Bank of the Russian Federation.

The draft decision on energy risk insurance has provided for one point. It is noted that if there is a shortage of funds in the insurance fund, state credit products are used in insurance cases. At the same time, there was a request to the Government to provide initial contributions to the funds being formed. Just all these steps became the most important stumbling block, which lasted for several more years.

At the same time, in May 1994, on the 16th, the European Parliament adopted Directive No. 94/19 regulating the deposit guarantee system. A special feature of the compulsory insurance system is the participation of all banks in such a system, which thus becomes insurance members. The same system provides the same guarantees for projects of different banks.

However, even if all these advantages are considered, it is worth highlighting the weakening motivation, as well as the weakening desire of customers to search for a reliable bank. At the same time, banks 'costs increase due to payments to insurance funds, which ultimately increases the cost of services provided by banks. Such systems operate in Japan, Finland, as well as in the USA, Canada, and other countries.

As for the voluntary system, if it exists, banks have the right to participate in such a system or to opt out of it. Those banks that do not participate in energy risk insurance are less competitive in the market of products provided by banks. The lack of competition is related to the fact that customers treat banks that are not participants in energy risk insurance with less confidence, but risk making a deposit with a forecast for a higher level of income.

For this reason, it is necessary to attract clients to banks without guarantees using the most common method, namely by raising rates, which is useful for customers, but not useful for the bank, whose costs inevitably increase. Even if the membership is voluntary, the authorities regulating the activity of banks and the possibility of joining still provide for certain restrictions on banking activities.

As an example, regulatory authorities may request a bank to provide insurance coverage, without which the bank is not granted a license to conduct certain operations. Also, if a bank does not have insurance coverage, it may not become a member of the association of banks. The state does all these actions to encourage banks to join the insurance system based on their own decision. The next classification is the amount of energy risk insurance guarantees. Here there are full size, limited, and discretionary sizes. Complete systems provide a reliable guarantee of payment on deposits, which also indicates the growing confidence of customers in banks.

Limited systems are defined as guaranteed provision of only partial coverage of customer energy risks. Most often, such guarantees are given only to small projects that are not very well oriented in the market environment. Naturally, large customers still have the motivation to choose the right banking institution.

5 Conclusions and Discussion

For commercial banking institutions, such a system makes it much easier to work with various small clients, and due to reduced fees, possible costs are reduced, while prices for services will not increase. However, this system has a small difficulty in determining the optimal amount of the amount that is subject to insurance. The following system, discretionary, is one of the types of limited system in which the insurance object expands during a crisis of the entire banking system. This system is the most flexible among the others.

Another classification is the degree of State participation. According to this classification, there are such types of insurance systems as public, private, and mixed. Most often, state systems are formed in the process of maintaining a system of mandatory energy risk insurance. In this case, insurance organizations are formed as state-owned, operating on a non-commercial basis.

The resources of this company are formed using state financial resources, as well as using contributions from banking institutions. This form is used to create insurance systems in the UK and the USA. Private energy risk insurances are formed and implemented at the expense of specialized organizations, whose activities are financed by financial contributions from participating banks. In this case, the state does not interfere in any way in these processes. Private energy risk insurances exist in Germany, France, and Luxembourg.

Finally, in mixed energy risk insurance, banks and the state are equally involved in creating resources for the insurance company. One example of such a mixed system is Japan, where the authorized capital has been operating since 1971, and it was created by the government, the state bank, and private banks, and in equal shares.

The last classification is the organization of financing cash payments or the method of accumulating funds. In this case, the systems may or may not have funding. Systems with financing are since specialized funds are created for payments of deductions for insurance. Funds are formed using regular contributions made by participating banks. Such a system has a fruitful effect on increasing trust, and in the event of an insured event, such a system also accelerates the transfer of funds as compensation. As for the system without financing, here the funds needed for compensation can only be found if necessary, such as the bankruptcy of a banking institution. This is a less preferable system, because in the event of a crisis, many banks come under attack, and it is very difficult to collect the necessary amount. Also, fundraising in this system is a long process, which causes panic among numerous projects. The goals of energy risk insurance based on the rapid elimination of the crisis and its consequences, as well as the formation of a stable system, cannot be achieved (Cheng et al., 2020; Haiyun et al., 2021; Zhe et al., 2021).

References

Alwaelya, S. A., Yousif, N. B. A., & Mikhaylov, A. (2021). Emotional development in preschoolers and socialization. *Early Child Development and Care, 190*(1717480), 2484–2493.

An, J., & Mikhaylov, A. (2020). Russian energy projects in South Africa. *Journal of Energy in Southern Africa, 31*(3), 58–64.

An, J., & Mikhaylov, A. (2021). Analysis of energy projects financial efficiency and renewable energy generation in Russia. *Finance: Theory and Practice, 25*(5), 1–11.

An, J., Mikhaylov, A., & Sokolinskaya, N. (2019a). Machine learning in economic planning: Ensembles of algorithms. *Journal of Physics: Conference Series, 1353*, 012126.

An, J., Mikhaylov, A., & Moiseev, N. (2019b). Oil price predictors: Machine learning approach. *International Journal of Energy Economics and Policy, 9*(5), 1–6.

An, J., Mikhaylov, A., & Richter, U. H. (2020a). Trade war effects: Evidence from sectors of energy and resources in Africa. *Heliyon, 6*, e05693. https://doi.org/10.1016/j.heliyon.2020.e05693

An, J., Mikhaylov, A., & Jung, S.-U. (2020b). The strategy of South Korea in the global oil market. *Energies, 13*(10), 2491.

An, J., Mikhaylov, A., & Kim, K. (2020c). Machine learning approach in heterogeneous group of algorithms for transport safety-critical system. *Applied Sciences, 10*(8), 2670.

An, J., Mikhaylov, A., & Jung, S.-U. (2021). A linear programming approach for robust network revenue management in the airline industry. *Journal of Air Transport Management, 91*, 101979.

Bhuiyan, M. A., An, J., Mikhaylov, A., Moiseev, N., & Danish, M. S. S. (2021). Renewable energy deployment and COVID-19 measures for sustainable development. *Sustainability, 13*, 4418. https://doi.org/10.3390/su13084418

Candila, V., Maximov, D., Mikhaylov, A., Moiseev, N., Senjyu, T., & Tryndina, N. (2021). On the relationship between oil and exchange rates of oil-exporting and oil-importing countries: From the great recession period to the Covid-19 era. *Energies, 14*, 8065.

Cheng, F., Lin, M., Yüksel, S., Dinçer, H., & Kalkavan, H. (2020). A hybrid hesitant 2-tuple IVSF decision making approach to analyze PERT-based critical paths of new service development process for renewable energy investment projects. *IEEE Access, 9*, 3947–3969.

Conteh, F., Takahashi, H., Hemeida, A. M., Krishnan, N., Mikhaylov, A., & Senjyu, T. (2021). Hybrid grid-connected renewable power generation for sustainable electricity supply in Sierra Leone: Case study lungi town. *Sustainability, 13*(18), 11435.

Danish, M. S. S., Bhattacharya, A., Stepanova, D., Mikhaylov, A., Grilli, M. L., Khosravy, M., & Senjyu, T. (2020). A systematic review of metal oxide applications for energy and environmental sustainability. *Metals, 10*(12), 1604.

Danish, M. S. S., Senjyu, T., Sabory, N. R., Khosravy, M., Grilli, M. L., Mikhaylov, A., & Majidi, H. (2021). A forefront framework for sustainable aquaponics Modeling and design. *Sustainability, 13*, 9313. https://doi.org/10.3390/su13169313

Dayong, N., Mikhaylov, A., Bratanovsky, S., Shaikh, Z. A., & Stepanova, D. (2020). Mathematical modeling of the technological processes of catering products production. *Journal of Food Process Engineering, 43*(2), 1–9.

Denisova, V., Mikhaylov, A., & Lopatin, E. (2019). Blockchain infrastructure and growth of global power consumption. *International Journal of Energy Economics and Policy, 9*(4), 22–29.

Dong, B., Ikonnikova, I., Rogulin, R., Sakulyeva, T., & Mikhaylov, A. (2021). Environmental-economic approach to optimization of transport communication in megacities. *Journal of Environmental Science and Health, Part A, 56*(6), 660–666. https://doi.org/10.1080/10934529.2021.1913928

Dooyum, U. D., Mikhaylov, A., & Varyash, I. (2020). Energy security concept in Russia and South Korea. *International Journal of Energy Economics and Policy, 10*(4), 102–107.

Dorofeev, M. L. (2020). Impact of monetary policy on the level of economic inequality in the United States. *MGIMO Review of International Relations, 13*(5), 97–114. https://doi.org/10.24833/2071-8160-2020-5-74-97-114

Fang, S., Zhou, P., Dinçer, H., & Yüksel, S. (2021). Assessment of safety management system on energy investment risk using house of quality based on hybrid stochastic interval-valued intuitionistic fuzzy decision-making approach. *Safety Science, 141*, 105333.

Grilli, M. L., Valerini, D., Rizzo, A., Yilmaz, M., Song, C., Hu, G., Chierchia, R., Mikhaylov, A., & Rinaldi, A. (2021). A comparative study of the mechanical and tribological properties of thin Al2O3 coatings fabricated by atomic layer deposition and radiofrequency sputtering. *Physica Status Solidi (A), 218*(24), 1–25.

Gura, D., Mikhaylov, A., Glushkov, S., Zaikov, M., & Shaikh, Z. A. (2020). Model for estimating power dissipation along the interconnect length in single on-chip topology. *Evolutionary Intelligence*, s12065. https://doi.org/10.1007/s12065-020-00407-7

Haiyun, C., Zhixiong, H., Yüksel, S., & Dinçer, H. (2021). Analysis of the innovation strategies for green supply chain management in the energy industry using the QFD-based hybrid interval valued intuitionistic fuzzy decision approach. *Renewable and Sustainable Energy Reviews, 143*, 110844.

Huang, Y., Masrur, H., Shigenobu, R., Hemeida, A. M., Mikhaylov, A., & Senjyu, T. (2021a). A comparative design of a campus microgrid considering a multi-scenario and multi-objective approach. *Energies, 14*, 2853. https://doi.org/10.3390/en14112853

Huang, Y., Yona, A., Takahashi, H., Hemeida, A. M., Mandal, P., Mikhaylov, A., Senjyu, T., & Lotfy, M. E. (2021b). Energy management system optimization of drug store operate electric vehicles Charging Station. *Sustainability, 14*, 1217836.

Ivanyuk, V. (2018). Econometric forecasting models based on forecast combination methods. In: *Proceedings of 2018 11th International Conference Management of Large-Scale System Development*, MLSD 2018. https://doi.org/10.1109/MLSD.2018.8551825.

Ivanyuk, V., & Berzin, D. (2020). An algorithm for constructing an efficient investment portfolio. In R. Silhavy, P. Silhavy, & Z. Prokopova (Eds.), *Software engineering perspectives in intelligent systems. CoMeSySo 2020. Advances in intelligent systems and computing* (Vol. 1294). Springer. https://doi.org/10.1007/978-3-030-63322-6_39

Ivanyuk, V., & Levchenko, K. (2020). Intelligent methods for predicting financial time series. In R. Silhavy, P. Silhavy, & Z. Prokopova (Eds.), *Software engineering perspectives in intelligent systems. CoMeSySo 2020. Advances in intelligent systems and computing* (Vol. 1294). Springer. https://doi.org/10.1007/978-3-030-63322-6_41

Ivanyuk, V., Soloviev, V. (2019). Efficiency of neural networks in forecasting problems. In: *Proceedings of 2019 12th International Conference Management of Large-Scale System Development*; MLSD 2019. https://doi.org/10.1109/MLSD.2019.8911046.

Ivanyuk, V., Sunchalin, A., & Sunchalina, A. (2020). Development of an intelligent ensemble forecasting system. In R. Silhavy, P. Silhavy, & Z. Prokopova (Eds.), *Software engineering perspectives in intelligent systems. CoMeSySo 2020. Advances in intelligent systems and computing* (Vol. 1294). Springer. https://doi.org/10.1007/978-3-030-63322-6_40

Jun, Q., Dinçer, H., & Yüksel, S. (2021). Stochastic hybrid decision-making based on interval type 2 fuzzy sets for measuring the innovation capacities of financial institutions. *International Journal of Finance & Economics, 26*(1), 573–593.

Kou, G., Akdeniz, Ö. O., Dinçer, H., & Yüksel, S. (2021). Fintech investments in European banks: A hybrid IT2 fuzzy multidimensional decision-making approach. *Financial Innovation, 7*(1), 1–28.

Li, X., Zhu, S., Yüksel, S., Dinçer, H., & Ubay, G. G. (2020). Kano-based mapping of innovation strategies for renewable energy alternatives using hybrid interval type-2 fuzzy decision-making approach. *Energy, 211*, 118679.

Lisin, A. (2020). Valuation of the activities of foreign banks in the Russian banking sector. *Orbis, 15*(45), 53–63. http://www.revistaorbis.org.ve/pdf/45/art5.pdf

Liu, Z.-J., Panfilova, E., Mikhaylov, A., & Kurilova, A. (2021a). COVID-19 crisis impact on the stability between parties in crowdfunding and crowdsourcing. *Wireless Personal Communications, 122*, 915–930. https://doi.org/10.1007/s11277-021-08932-z

Liu, Y., Gong, X., Yüksel, S., Dinçer, H., & Aydın, R. (2021b). A multidimensional outlook to energy investments for the countries with continental shelf in East Mediterranean region with hybrid decision making model based on IVIF logic. *Energy Reports, 7*, 158–173.

Liu, Z.-J., Panfilova, E., Mikhaylov, A., & Kurilova, A. (2022). Assessing stability in relationship between parties in crowdfunding and crowdsourcing projects during the COVID-19 crisis. *Journal of Global Information Management, 30*(2), 1.

Melnichuk, A. V., Melnichuk, Y. A., Bocharov, A. V., Stepanova, D. I., & Kurbanov, S. A. (2020). Constitutional foundations of the functioning of local self-government. *Revista Inclusiones., 7*, 337–346.

Meynkhard, A. (2019). Fair market value of bitcoin: Halving effect. *Investment Management and Financial Innovations, 16*(4), 72–85. https://doi.org/10.21511/imfi.16(4).2019.07

Meynkhard, A. (2020). Effect of bitcoin volatility on altcoins pricing. *Advances in Intelligent Systems and Computing, 1294*, 652–664. https://doi.org/10.1007/978-3-030-63322-6_55

Mikhaylov, A. (2018a). Volatility spillover effect between stock and exchange rate in oil exporting countries. *International Journal of Energy Economics and Policy, 8*(3), 321–326.

Mikhaylov, A. (2018b). Pricing in oil market and using probit model for analysis of stock market effects. *International Journal of Energy Economics and Policy, 8*(2), 69–73.

Mikhaylov, A. (2018c). The yield of the carry trade strategy. *Finance: Theory and Practice, 22*(3), 52–63. https://doi.org/10.26794/2587-5671-2018-22-3-52-63

Mikhaylov, A. (2020a). Cryptocurrency market analysis from the open innovation perspective. *Journal of Open Innovation: Technology, Market, and Complexity, 6*(4), 197. https://doi.org/10.3390/joitmc6040197

Mikhaylov, A. (2020b). Geothermal energy development in Iceland. *International Journal of Energy Economics and Policy, 10*(4), 31–35.

Mikhaylov, A. (2020c). Cryptocurrency market development: Hurst method. *Finance: Theory and Practice, 24*(3), 81–91.

Mikhaylov, A. (2021a). Development of Friedrich von Hayek's theory of private money and economic implications for digital currencies. *Terra Economicus, 19*(1), 53–62.

Mikhaylov, A. (2021b). Lichens as indicators of atmospheric pollution in urban ecosystems. *Israel Journal of Ecology & Evolution, 10016*, 1–9.

Mikhaylov, A. (2022). Efficiency of renewable energy generation in Russia. *Annals of the Brazilian Academy of Sciences, 94*(3), 1.

Mikhaylov, A., & Tarakanov, S. (2020). Development of Levenberg-Marquardt theoretical approach for electric network. *Journal of Physics: Conference Series, 1515*, 052006.

Mikhaylov, A., Sokolinskaya, N., & Nyangarika, A. (2018). Optimal carry trade strategy based on currencies of energy and developed economies. *Journal of Reviews on Global Economics, 7*, 582–592.

Mikhaylov, A., Burakov, D., & Didenko, V. (2019). Relationship between oil Price and macroeconomic indicators in Russia. *Finance: Theory and Practice, 23*(2), 105–116. https://doi.org/10.26794/2587-5671-2019-23-2-105-116

Mikhaylov, A., Yumashev, A., & Kolpak, E. (2021a). Quality of life, anxiety and depressive disorders in patients with extrasystolic arrhythmia. *Archives of Medical Science, 17*(1). https://doi.org/10.5114/aoms.2020.101359

Mikhaylov, A., Danish, M. S. S., & Senjyu, T. (2021b). New stage in evolution of cryptocurrency market: Analysis by Hurst method. In H. Dincer & S. Yuksel (Eds.), *Strategic outlook in business and finance innovation: Multidimensional policies for emerging economies* (pp. 35–45. ISBN: 978-1-80043-445-5). Emerald Publishing Limited. https://doi.org/10.1108/978-1-80043-444-820211004

Moiseev, N., Mikhaylov, A., Varyash, I., & Saqib, A. (2020). Investigating the relation of GDP per capita and corruption index. *Entrepreneurship and Sustainability Issues, 8*(1), 780–794.

Moiseev, N., Sorokin, A., Zvezdina, N., Mikhaylov, A., Khomyakova, L., & Danish, M. S. S. (2021). Credit risk theoretical model on the base of DCC-GARCH in time-varying parameters framework. *Mathematics, 9*, 2423. https://doi.org/10.3390/math9192423

Morkovkin, D. E., Gibadullin, A. A., Kolosova, E. V., Semkina, N. S., & Fasehzoda, I. S. (2020a). Modern transformation of the production base in the conditions of industry 4.0: Problems and prospects. *Journal of Physics: Conference Series, 1515*, 032014.

Morkovkin, D., Lopatkin, D., Sadriddinov, M., Shushunova, T., Gibadullin, A., & Golikova, O. (2020b). Assessment of innovation activity in the countries of the world. *E3S Web of Conferences, 157*, 04015. https://doi.org/10.1051/e3sconf/202015704015

Mukhametov, A., Bekhorashvili, A., Avdeenko, A., & Mikhaylov, A. (2021). The impact of growing legume plants under conditions of Biologization and soil cultivation on chernozem fertility and productivity of rotation crops. *Legume Research, 44*, 12.

Mutalimov, V., Kovaleva, I., Mikhaylov, A., & Stepanova, D. (2021). Assessing regional growth of small business in Russia. *Entrepreneurial Business and Economics Review, 9*(3), 119–133. https://doi.org/10.15678/EBER.2021.090308

Nie, D., Panfilova, E., Samusenkov, V., & Mikhaylov, A. (2020). E-learning financing models in Russia for sustainable development. *Sustainability, 12*(11), 4412.

Nyangarika, A., Mikhaylov, A., & Tang, B.-J. (2018). Correlation of oil prices and gross domestic product in oil producing countries. *International Journal of Energy Economics and Policy, 8*(5), 42–48.

Nyangarika, A., Mikhaylov, A., & Richter, U. (2019a). Influence oil price towards economic indicators in Russia. *International Journal of Energy Economics and Policy, 9*(1), 123–129.

Nyangarika, A., Mikhaylov, A., & Richter, U. (2019b). Oil price factors: Forecasting on the base of modified auto-regressive integrated moving average model. *International Journal of Energy Economics and Policy, 9*(1), 149–159.

Qiu, D., Dinçer, H., Yüksel, S., & Ubay, G. G. (2020). Multi-faceted analysis of systematic risk-based wind energy investment decisions in E7 economies using modified hybrid modeling with IT2 fuzzy sets. *Energies, 13*(6), 1423.

Radosteva, M., Soloviev, V., Ivanyuk, V., & Tsvirkun, A. (2018). Use of neural network models in market risk management. *Advances in Systems Science and Applications, 18*(2), 53–58. https://doi.org/10.25728/assa.2018.18.2.582

Ranjbar, O., Chang, T., Nel, E., & Gupta, R. (2017). Energy consumption and economic growth nexus in South Africa: Asymmetric frequency domain approach. *Energy Sources, Part B: Economics, Planning, and Policy, 12*(1), 24–31.

Rathnayaka, R. K. T., Seneviratna, D. M. K., & Long, W. (2018). The dynamic relationship between energy consumption and economic growth in China. *Energy Sources, Part B: Economics, Planning, and Policy, 13*(5), 264–268.

Saqib, A., Chan, T.-H., Mikhaylov, A., & Lean, H. H. (2021). Are the responses of sectoral energy imports asymmetric to exchange rate volatilities in Pakistan? Evidence from recent foreign exchange regime. *Frontiers in Energy Research, 9*, 531. https://doi.org/10.3389/fenrg.2021.614463

Silahtaroğlu, G., Dinçer, H., & Yüksel, S. (2021). Defining the significant factors of currency exchange rate risk by considering text mining and fuzzy AHP. In *Data science and multiple criteria decision making approaches in finance* (pp. 145–168). Springer.

Sunchalin, A. M., Kochkarov, R. A., Levchenko, K. G., Kochkarov, A. A., & Ivanyuk, V. A. (2019). Methods of risk management in portfolio theory. *Espacios, 40*(16), 8.

Uandykova, M., Lisin, A., Stepanova, D., Baitenova, L., Mutaliyeva, L., Yuksel, S., & Dincer, H. (2020). The social and legislative principles of counteracting ransomware crime. *Entrepreneurship and Sustainability Issues, 8*(2), 777–798.

Udalov, I. (2021). The transition to renewable energy sources as a threat to resource economies. *International Journal of Energy Economics and Policy, 11*(3), 1–6.

Uyeh, D. D., Asem-Hiablie, S., Park, T., Kim, K.-M., Mikhaylov, A., Woo, S., & Ha, Y. (2021). Could japonica Rice be an alternative variety for increased global food security and climate change mitigation? *Food, 10*, 1869. https://doi.org/10.3390/foods10081869

Varyash, I., Mikhaylov, A., Moiseev, N., & Aleshin, K. (2020). Triple bottom line and corporate social responsibility performance indicators for Russian companies. *Entrepreneurship and Sustainability Issues, 8*(1), 313–331.

Yuan, G., Xie, F., Dinçer, H., & Yüksel, S. (2021). The theory of inventive problem solving (TRIZ)-based strategic mapping of green nuclear energy investments with spherical fuzzy group decision-making approach. *International Journal of Energy Research, 45*(8), 12284–12300.

Yüksel, S., Khomyakova, L., & Mikhaylov, A. (2021a). Energy Center selection in G7 industry with fuzzy MOORA. In S. Yüksel & H. Dinçer (Eds.), *Handbook of research on strategic Management for Current Energy Investments* (pp. 87–106). IGI Global. ISBN13: 9781799883357. https://doi.org/10.4018/978-1-7998-8335-7.ch006

Yüksel, S., Mikhaylov, A., Ubay, G. G., & Uyeh, D. D. (2021b). The role of hydrogen in the Black Sea for the future energy supply security of Turkey. In S. Yüksel & H. Dinçer (Eds.), *Handbook of research on strategic Management for Current Energy Investments* (pp. 1–15). IGI Global. https://doi.org/10.4018/978-1-7998-8335-7.ch001

Yüksel, S., Mikhaylov, A., & Ubay, G. G. (2021c). Factors causing delay in the installation of nuclear power plants. In S. Yüksel & H. Dinçer (Eds.), *Strategic approaches to energy management. Contributions to management science*. Springer. https://doi.org/10.1007/978-3-030-76783-9_7

Yumashev, A., & Mikhaylov, A. (2020). Development of polymer film coatings with high adhesion to steel alloys and high wear resistance. *Polymer Composites, 41*(7), 2875–2880.

Yumashev, A., Ślusarczyk, B., Kondrashev, S., & Mikhaylov, A. (2020). Global indicators of sustainable development: Evaluation of the influence of the human development index on consumption and quality of energy. *Energies, 13*, 2768.

Zhao, Q., Cherkasov, S., Avdeenko, A., Kondratenko, L., & Mikhaylov, A. (2021). Integral estimate of the added value of agricultural crops in the synergetic agriculture on the example of vicia faba (Vicia faba L.). *Polish Journal of Environmental Studies, 30*(5), 1–9. https://doi.org/10.15244/pjoes/131949

Zhe, L., Yüksel, S., Dinçer, H., Mukhtarov, S., & Azizov, M. (2021). The positive influences of renewable energy consumption on financial development and economic growth. *SAGE Open, 11*(3), 21582440211040133.

Zhou, P., Luo, J., Cheng, F., Yüksel, S., & Dinçer, H. (2021). Analysis of risk priorities for renewable energy investment projects using a hybrid IT2 hesitant fuzzy decision-making approach with alpha cuts. *Energy, 224*, 120184.

ICT Trade and Energy Transition in the BRICS Economies

Ibrahim Nandom Yakubu, Ayhan Kapusuzoglu, and Nildag Basak Ceylan

1 Introduction

Energy is increasingly recognized as a key component in the production process and its demand is growing exponentially globally. The traditional energy sources such as coal, oil, and natural gas (all of which are classified as non-renewable forms of energy) have a strong influence on the growth and prosperity of most economies (Ellabban et al., 2014). These energy sources are also accessible and continue to provide good options for powering automobiles. Despite the benefits of non-renewable energy sources, they have several drawbacks. Among the problems associated with the use of non-renewable energy sources is the increasing emission of carbon dioxide (CO_2), which primarily contributes to climate change.

Given the downsides of the natural energy sources, switching from the need for non-renewables towards the usage of renewable energy has been massively advocated, and most countries have responded to these campaigns by gradually moving their emphasis to these renewable sources (Asiedu et al., 2021). As per the International Energy Outlook, renewable energy demand has accelerated globally over the years with an anticipation that it will reach 50% by the year 2050.

Among the world regional blocs, the BRICS countries constituted by Brazil, Russia, India, China, and South Africa have experienced a rapid transformation with growth in the level of economic activities (Pathak & Shah, 2019). In the energy landscape, the BRICS bloc is also among the leading suppliers and consumers of energy in the world. To cite, the International Energy Agency (IEA) reported that the

I. N. Yakubu
Graduate School of Social Sciences, Ankara Yildirim Beyazit University, Ankara, Turkey

A. Kapusuzoglu (✉) · N. B. Ceylan
Faculty of Business, Ankara Yildirim Beyazit University, Ankara, Turkey
e-mail: akapusuzoglu@ybu.edu.tr; nbceylan@ybu.edu.tr

© The Author(s), under exclusive license to Springer Nature Switzerland AG 2022
H. Dinçer, S. Yüksel (eds.), *Sustainability in Energy Business and Finance*, Contributions to Finance and Accounting,
https://doi.org/10.1007/978-3-030-94051-5_2

bloc's contribution to the overall energy supply in the year 2017 was estimated at 36.4%, placing it as the second-largest energy supplier after the OECD.

Aside from the contribution to the non-renewable energy base, the BRICS alliance is also emerging as one of the leading participants in the green energy sector, as some members in the bloc are rapidly substituting "dirty energy" sources with renewable energy. Per the United Nations (UN) report issued in the year 2018, China alone accounted for about 45% of the entire sum of green energy investment on a global scale. The report further established that the renewable energy investment by China, Brazil, and India labeled as the "Big Three" in the BRICS bloc amounted to over half of the total renewable energy investment globally. Considering the drive to achieving clean energy by the BRICS economies, it is of importance to investigate what factors influence the process of acclimating from non-renewables to renewable energy sources in the bloc. Hence, the aim of this study. The literature has outlined several factors as drivers of renewable energy consumption. Among these factors include economic growth (Alam & Murad, 2020), foreign direct investment (Polat, 2018), trade openness (Murshed, 2018a, 2018b), institutional factors (Chen et al., 2021), and many others.

Notwithstanding the calls for commitment for clean energy, technological impoverishment is commonly cited as a key roadblock to accomplishing the energy transition targets (Murshed et al., 2020). The trade of ICT goods is however expected to remove these technological constraints that have typically hampered the smooth energy transition processes of emerging economics. In this study, we investigate how ICT trade influences renewable energy transition employing BRICS economies. Further to exploring the possible impact of ICT trade on BRICS's energy transition, we examine whether the increase in openness to ICT trade simultaneously decreases CO_2 emission in the bloc. As far as we are aware, this research presents an initial attempt to address the influence of ICT trade on renewable energy transition with a focus on the BRICS countries.

The rest of this chapter is laid out as follows. The second part discusses the literature review. The method is given in Sect. 3. The fourth part highlights the findings and discussions, while Sect. 5 provides the conclusions and policy suggestions.

2 Literature Review

This part of the chapter is divided into different sub-headings, the first of which examines the theoretical basis of the study, and the second of which sheds light on the related empirical findings presented in the literature on ICT, trade-renewable energy linkage and ICT, trade-environmental pollution nexus.

2.1 Theoretical Background

The Heckscher-Ohlin model, which was further developed by Vanek (1968) can be used to comprehend the rational basis for easing barriers to trade. The theory illustrates how trade activities are carried out, particularly across countries with disparate features. The theory argues that countries with a high concentration of factor endowment receive a significant amount of investment from abroad (Ohlin 1933). Therefore, a country will strive to be a leading exporter of commodities that heavily utilize its bountiful factors while importing items that profoundly utilize its scarce resources. Following this theory, the removal of trade restrictions will accelerate the flows and transfer of ICT goods to locations or countries with substantially lower factor endowments in the production of these goods given their available resources. Increasing ICT trade openness will contribute to the creation of a solid ICT infrastructure which is expected to facilitate a smooth energy transition process through the successful implementation of renewable energy technologies.

2.2 ICT, Trade Openness-Renewable Energy Nexus

ICT trade and the use of renewable energy resources are not well studied in the literature. In spite of this, numerous studies have documented how increasing trade openness drives the usage of renewable energy, particularly in emerging markets. Given the notion that ICT products trading volume forms a fraction of the entire trade amounts of nations, the extant studies on the trade openness-renewable energy adoption can give a better idea and the perception on the link between ICT trade and the issue of renewable energy demand.

On the trade openness renewable energy consumption nexus, Murshed (2018a, 2018b) looked into the impact of trade openness on the energy transition process of some selected Asian countries over the period 2000–2017. Applying the two-stage least squares technique, the results demonstrated that a boost in trade increases the use of renewable energy in the studied countries. Using a sample of 25 countries in the OECD bloc, Alam and Murad (2020) analyzed how trade openness and some other factors facilitate renewable energy consumption. Employing different panel estimation methods, the authors revealed that renewable energy consumption is significantly triggered by an increase in trade openness. In the instance of Malaysia, Lau et al. (2018) examined the factors influencing renewable energy usage for the years 1980–2015. The findings showed that in the long term, trade openness negatively motivates the consumption of renewable electricity. Uzar (2020), in a cross-country study involving both advanced and developing countries, examined the factors driving renewable energy consumption. The conclusions of the ARDL approach revealed that trade liberalization had no considerable influence on renewable energy use. Employing data of countries from Sub-Saharan Africa, Asia, Latin America, and the Caribbean Islands, Murshed (2018a, 2018b) noted that enhanced trade on average inhibits renewable energy utilization. Using yearly data spanned 1971–2015, Shakouri and Khoshnevis Yazdi (2017) analyzed the correlation between openness to trade and energy usage in South Africa. With the ARDL approach, the researchers evidenced that the variation in energy demand is influenced by trade openness, and a two-way interaction exists between trade openness and demand for renewable energy. Basu et al. (2020) explored the effect of trade openness and other factors on the share of renewable energy in India. The study reported that a surge in trade facilitates the implementation of renewable and energy-efficient technologies. Applying the vector error correction model, Lin et al. (2016) scrutinized the motivators of renewable electricity demand in the case of China. Evidence from the analysis showed that the level of openness to trade impedes renewable electricity demand. Using a sample of South Asian countries, Murshed (2020) examined the impact of ICT trade on energy transition. The findings of the study depicted that ICT trade boosts the usage of renewable energy while simultaneously increasing the percentage of renewable energy in the final energy demand. Wang and Zhang (2021) analyzed how free trade affects renewable energy using a sample of 186 countries across the globe. The study suggested a direct influence of free trade on renewable energy in economies classified as high- and upper-middle-income, albeit with an inverse impact on lower-middle-income nations.

Notwithstanding the deficient empirical studies on the influence of ICT trade on renewable energy transition, some studies have acknowledged the significance of ICT infrastructure in leveraging the uptake of renewable energy sources. Stallo et al. (2010), for example, opined that the adoption of sophisticated ICT products can augment existing processes to enhance power generation from renewable energy sources. Ahmed et al. (2017) affirmed that the growth in the ICT sector facilitates energy transition through the use of ICT products which aids in energy conservation.

2.3 ICT, Trade Openness-Carbon Dioxide (CO_2) Emission Nexus

Given the second objective which is to examine how ICT trade contributes to CO_2 emission, we review the literature on the influence of trade openness and ICT on CO_2 emission. For the impact of trade on emission levels, Managi et al. (2009) assessed the environmental outcome of trade openness using data of developing and developed economies over the period 1973–2000. The authors showed that trade improves environmental quality in OECD economies and increases CO_2 emission in countries that are not in the OECD region. Similarly, in the OECD countries, Gozgor (2017) posited a long-term negative influence of trade openness on CO_2 emission. Li and Qi (2011) analyzed the trade openness and CO_2 emission link in the case of China for the years 1997–2008. Applying different estimation approaches, the results established that CO_2 emission upsurges with improvement in trade openness. Shahbaz et al. (2017) assessed the influence of trade on the emission of CO_2 using a panel of 105 economies. The results from the vector error correlation model indicated that overall, openness to trade has a detrimental impact on the environment though with varying effects among country classifications. Employing a sample of emerging economies with data that spanned from 1990–2013, Saidi and Mbarek (2017) explored the effect trade openness has on CO_2 emission. The results showed no significant impact of trade on emissions levels in the countries examined. Using the pooled mean group technique, Park et al. (2018) found that increasing trade levels lessens CO_2 emission in the European Union. In the Belt and Road region, Sun et al. (2019) discovered that trade openness leads to an increasing level of CO_2 emission. In investing the impact of openness to trade on pollution levels, Tachie et al. (2020) invoked the mean group approach on data collected from the EU-18. The results reported that CO_2 emission magnifies with growth in trade. Ali et al. (2021) evidenced that trade coupled with technological innovation mitigates the emission of CO_2 while trade stimulated by economic growth increases pollution levels in Asian countries. In analyzing the long-term link between trade and CO_2 emission, Sun et al. (2020) applied the panel cointegration techniques with data from Sub-Saharan African countries. The researchers established that CO_2 emission in the long term is somehow reduced as trade increases.

The extant literature has also documented how the use of ICT influences CO_2 emissions though with conflicting findings. For instance, relying on the STIRPAT model, Zhang and Liu (2015) established that the growth of ICT industry in China lowers the emissions of carbon dioxide. Raheem et al. (2020) noted that in the G7 alliance, ICT directly influences emissions in the long term. Nguyen et al. (2020) using G20 countries reported a converse relationship between ICT and emissions. Amri et al. (2019) observed no significant association between ICT and emissions levels in Tunisia. Invoking the Quantile ARDL method, Godil et al. (2020) established that emissions levels in Pakistan are negatively driven by ICT. Lu (2018) showed a negative significant influence of ICT on the level of emissions in 12 Asian economies. Similarly, according to Haini (2021), the use of ICT products in ASEAN economies continuously lessens the emission of carbon dioxide. Applying the ARDL model, Khanal (2021) revealed that ICT in the long run inimically affects the environment.

From the literature review, it can be noted that the direct influence of ICT trade on renewable energy transition and CO_2 emissions is virtually absent. However, ample studies have considered how trade in general influences renewable energy and emissions levels. Thus, this study intends to contribute to filling a void in the literature regarding the impact of ICT trade (a segment of net trade) on the energy transition process and CO_2 emissions in the case of the BRICS economies.

3 Research Methodology

3.1 Data and Variable Selection

To achieve the study's goal, we use data from the BRICS economies spanning the years 2000–2017. The data are gathered from the World Bank's World Development Indicators. Renewable energy consumption (as a percentage of total final energy consumption) is the primary dependent variable for our analysis. To gauge CO_2 emissions, we utilize CO_2 emissions in metric tons per capita following the proxy used by Yakubu et al. (2021a). ICT trade is measured as the percentage share of ICT goods imports and exports in total imports and exports of goods. The study takes into account the influence of foreign direct investment and economic growth, which serves as control variables. All parameters are converted into the natural logarithm form.

3.2 Model Specification

In line with Yakubu et al. (2021b), the following basic empirical models are specified to scrutinize the effect of the regressors on renewable energy demand and CO_2 emissions:

$$ln\,\mathrm{REN}_{it} = \alpha_0 + \beta_1\, ln\,\mathrm{CT\text{-}TR}_{it} + \beta_2\, ln\,\mathrm{FDI}_{it} + \beta_3\, ln\,\mathrm{GDP}_{it} + \varepsilon_{it} \tag{1}$$

$$ln\,\mathrm{CO}_{2it} = \alpha_0 + \beta_1\, ln\,\mathrm{CT\text{-}TR}_{it} + \beta_2\, ln\,\mathrm{FDI}_{it} + \beta_3\, ln\,\mathrm{GDP}_{it} + \varepsilon_{it} \tag{2}$$

where REN, CO_2, ICT-TR, FDI, and GDP represent renewable energy consumption, CO_2 emissions, ICT trade, foreign direct investment, and economic growth in a specific country i at period t, respectively. ε denotes the error term.

3.3 Analytical Strategy

The dynamic ordinary least squares (DOLS) method is used to estimate the models. This approach is suitable for modeling long-run relationships as well as handling endogeneity problems in panel analysis. Also, compared to other panel cointegration techniques, the DOLS technique has enhanced asymptotic properties regarding bias in variable estimation and standard errors (Funk, 2001).

4 Empirical Results and Discussions

4.1 Stationarity and Cointegration Tests

To determine whether or not the series are stationary, the unit root features of the selected variables are explored prior to the models' estimations—this process aids in minimizing erroneous regression. To examine the variables' unit root features, a number of approaches have been developed in the extant literature. The Fisher Augmented Dickey-Fuller (ADF) and Phillips–Perron (PP) panel unit root tests techniques are employed in this study to evaluate the series unit root status. The results of the tests are reported in Table 1. Regarding the ADF test, renewable energy consumption and ICT trade are stationary at levels. In addition to renewable energy and ICT trade, FDI also shows stationarity when the unit root is examined using the PP test. In all, carbon dioxide emission and GDP are stationary at first difference for both ADF and PP tests. Nonetheless, all the variables are stationary at first difference with the same methods of unit root tests. It is worth noting that the traditional OLS technique is not suitable for our model estimation given the difference in stationarity level of the variables as postulated by the results of the unit root tests. As a result, the DOLS approach proposed by Kao and Chiang (2001) is applied in this research.

Prior to estimating the model, it is essential to assess the long-run relationship amid the selected variables in the study. In doing so, we invoke the panel cointegration test by Kao (1999). The test result is also shown in Table 1. The cointegration test indicates a long-term association amid the selected variables, thus justifying the estimation of our model.

Table 1 Stationary and cointegration tests results

Unit root test				
	Level		First difference	
Variables	ADF-Fisher	PP-Fisher	ADF-Fisher	PP-Fisher
lnREN	15.857**	15.457**	13.746*	37.074***
$\ln CO_2$	11.300	12.179	23.571***	54.624***
lnICT-TR	20.351**	29.389***	42.181***	90.160***
lnFDI	11.558	20.405**	36.138***	84.256***
lnGDP	13.286	12.651	17.222*	21.852***
Kao cointegration test				
ADF			***t*-statistic**	**Prob.**
			−2.607	0.005

Note: *, **, and *** denote the level significance at 10%, 5%, and 1% respectively

4.2 Regression Results

Table 2 presents the regression estimation from the dynamic OLS (DOLS). From model 1, we estimate the long-run influence of ICT trade on renewable energy consumption. It is glaring from the analysis that ICT trade openness exerts a significant positive effect on the use of renewable energy. Precisely, on average, a percentage growth in openness to ICT trade increases the demand for renewable energy by 1.461%. Thus, it can be asserted that removing obstacles particularly tariffs on the trading of ICT goods would be optimal for integrating renewable energy resources into the energy grid of the BRICS alliance. In other words, allowing the trade of ICT goods will enable members in the BRICS to adopt sophisticated ICT products that can augment existing processes to enhance power generation from renewable energy sources. This result sync with the findings of Murshed (2020) who documented that ICT trade significantly promotes the renewable energy transition in South Asia.

For the control factors, we note that while foreign direct investment positively enhances the use of renewable energy, economic growth inversely affects the consumption of renewable energy. Nevertheless, the impact of these variables is insignificant.

From model 2, we observe that ICT trade aids in reducing environmental degradation by lowering the emissions of CO_2 as indicated by the negative relationship between the two variables. Specifically, a percentage increase in ICT trade reduces CO_2 emission by 1.262%. The finding suggests that liberalizing trade by mitigating bottlenecks to the transfer of ICT products will ensure environmental sustainability through the reliance on "green" technologies.

Our analysis also shows that FDI significantly contributes to reducing CO_2 emissions in the BRICS. As FDI surges by a percentage, CO_2 emission declines by 0.686%. The finding indicates that FDI encourages the use of energy-efficient technologies, which greatly lower energy consumption with a subsequent declining impact on pollution. This result agrees with prior studies (Chang & Huang, 2015;

Table 2 DOLS estimation results

Variables	Model 1 (Renewable energy)	Model 2 (CO_2 emission)
lnICT-TR	1.461**	−1.262***
	(0.017)	(0.006)
lnFDI	0.268	−0.686***
	(0.435)	(0.000)
lnGDP	−0.079	0.548***
	(0.608)	(0.000)
R^2	0.678	0.891
Adj. R^2	0.578	0.860

Notes: p–values are in parentheses; **, and *** denote the level significance at 5% and 1% respectively

Demena & Afesorgbor, 2020; Hao and Liu 2015; Huang et al., 2019; Zhu et al., 2016).

Finally, the results indicate that economic growth significantly magnifies emission levels in the BRICS. The plausible implication is that as economic activities upsurge, which includes industrial activities, the demand for energy also rises, leading to more emissions in the atmosphere. The result is similar to earlier empirical findings on the economic growth-emissions relationship (Chekouri et al., 2021; Hassoun et al., 2018; Kahia et al., 2019; Khan et al., 2020; Osobajo et al., 2020; Ridzuan et al., 2020).

5 Conclusion and Policy Recommendations

The quest to achieve a sustainable environment through transitioning from the reliance on conventional energy sources to green energy resources has gained policy attention globally. One of the means to realizing this objective is by lowering trade barriers for easy movement of environmentally friendly technologies. In this chapter, the researchers assessed how the boost in trade with respect to ICT goods accelerates renewable energy transition in the BRICS nations. Using the dynamic OLS technique with data covering from 2000–2017, our findings report that ICT trade openness exerts a significant positive influence on renewable energy utilization and a significant negative effect on emissions of CO_2. The results clearly postulate that increasing trade of ICT products promotes the demand for renewable energy and mitigates environmental pollution, which supports the clean energy agenda of the BRICS. We note that FDI aids in reducing CO_2 emission though the influence on renewable energy demand is insignificant. Economic growth is reported to aggravate environmental pollution in the bloc.

In light of the results, sound policy initiatives must be implemented to expedite the flows and transfer of ICT products in the BRICS countries. This will not only support a successful energy transition through the reliance on renewable energy but will also help in averting environmental pollution. In advancing this study, future research may consider examining how openness to the trade of ICT products can enhance energy transition efforts for each of the countries in the BRICS to ascertain any significant variability in the results.

References

Ahmed, F., Naeem, M., & Iqbal, M. (2017). ICT and renewable energy: A way forward to the next generation telecom base stations. *Telecommunication Systems, 64*(1), 43–56.

Alam, M. M., & Murad, M. W. (2020). The impacts of economic growth, trade openness and technological progress on renewable energy use in organization for economic co-operation and development countries. *Renewable Energy, 145*, 382–390.

Ali, U., Li, Y., Yánez Morales, V. P., & Hussain, B. (2021). Dynamics of international trade, technology innovation and environmental sustainability: Evidence from Asia by accounting for cross-sectional dependence. *Journal of Environmental Planning and Management, 64*(10), 1–22.

Amri, F., Zaied, Y. B., & Lahouel, B. B. (2019). ICT, total factor productivity, and carbon dioxide emissions in Tunisia. *Technological Forecasting and Social Change, 146*, 212–217.

Asiedu, B. A., Hassan, A. A., & Bein, M. A. (2021). Renewable energy, non-renewable energy, and economic growth: Evidence from 26 European countries. *Environmental Science and Pollution Research, 28*(9), 11119–11128.

Basu, S., Roy, M., & Pal, P. (2020). Exploring the impact of economic growth, trade openness and urbanization with evidence from a large developing economy of India towards a sustainable and practical energy policy. *Clean Technologies and Environmental Policy, 22*(4), 877–891.

Chang, S. C., & Huang, W. T. (2015). The effects of foreign direct investment and economic development on carbon dioxide emissions. In H. Van-Nam, K. Vladik, S. Songsak, & K. Suriya (Eds.), *Econometrics of risk, studies in computational intelligence* (pp. 483–496). Springer.

Chekouri, S. M., Chibi, A., & Benbouziane, M. (2021). Economic growth, carbon dioxide emissions and energy consumption in Algeria: A wavelet coherence approach. *World Journal of Science, Technology and Sustainable Development, 18*(2), 172–189.

Chen, C., Pinar, M., & Stengos, T. (2021). Determinants of renewable energy consumption: Importance of democratic institutions. *Renewable Energy, 179*, 75–83.

Demena, B. A., & Afesorgbor, S. K. (2020). The effect of FDI on environmental emissions: Evidence from a meta-analysis. *Energy Policy, 138*, 111192.

Ellabban, O., Abu-Rub, H., & Blaabjerg, F. (2014). Renewable energy resources: Current status, future prospects and their enabling technology. *Renewable and Sustainable Energy Reviews, 39*, 748–764.

Funk, M. (2001). Trade and international R&D spillovers among OECD countries. *Southern Economic Journal, 67*(3), 725–736.

Godil, D. I., Sharif, A., Agha, H., & Jermsittiparsert, K. (2020). The dynamic nonlinear influence of ICT, financial development, and institutional quality on CO_2 emission in Pakistan: New insights from QARDL approach. *Environmental Science and Pollution Research, 27*(19), 24190–24200.

Gozgor, G. (2017). Does trade matter for carbon emissions in OECD countries? Evidence from a new trade openness measure. *Environmental Science and Pollution Research, 24*(36), 27813–27821.

Haini, H. (2021). Examining the impact of ICT, human capital and carbon emissions: Evidence from the ASEAN economies. *International Economics, 166*, 116–125.

Hao, Y., & Liu, Y. M. (2015). Has the development of FDI and foreign trade contributed to China's CO_2 emissions? An empirical study with provincial panel data. *Natural Hazards, 76*(2), 1079–1091.

Hassoun, S. E. S., Mékidiche, M., & Guellil, M. S. (2018). Examining the connection amongst renewable energy, economic growth and carbon dioxide emissions in Algeria. *Ekoist: Journal of Econometrics and Statistics, 14*(29), 199–223.

Huang, Y., Chen, X., Zhu, H., Huang, C., & Tian, Z. (2019). The heterogeneous effects of FDI and foreign trade on CO_2 emissions: Evidence from China. *Mathematical Problems in Engineering, 2019*, 1–14. Article ID: 9612492.

Kahia, M., Jebli, M. B., & Belloumi, M. (2019). Analysis of the impact of renewable energy consumption and economic growth on carbon dioxide emissions in 12 MENA countries. *Clean Technologies and Environmental Policy, 21*(4), 871–885.

Kao, C. (1999). Spurious regression and residual-based tests for cointegration in panel data. *Journal of Econometrics, 90*(1), 1–44.

Kao, C., & Chiang, M. H. (2001). On the estimation and inference of a cointegrated regression in panel data. In B. H. Baltagi, T. B. Fomby, & R. C. Hill (Eds.), *Non stationary panels, panel cointegration, and dynamic panels. (advances in econometrics)* (Vol. 15, pp. 179–222). Emerald Group Publishing Limited.

Khan, M. K., Khan, M. I., & Rehan, M. (2020). The relationship between energy consumption, economic growth and carbon dioxide emissions in Pakistan. *Financial Innovation, 6*(1), 1–13.

Khanal, A. (2021). The role of ICT and energy consumption on carbon emissions: An Australian evidence using cointegration test and ARDL long-run and short-run methodology. *International Journal of Energy Economics and Policy, 11*(5), 441–449.

Lau, L. S., Yii, K. J., Lee, C. Y., Chong, Y. L., & Lee, E. H. (2018). Investigating the determinants of renewable energy consumption in Malaysia: An ARDL approach. *International Journal of Business and Society, 19*(3), 886–903.

Li, K., & Qi, S. Z. (2011). Trade openness, economic growth and carbon dioxide emission in China. *Economic Research Journal, 11*(4), 60–72.

Lin, B., Omoju, O. E., & Okonkwo, J. U. (2016). Factors influencing renewable electricity consumption in China. *Renewable and Sustainable Energy Reviews, 55*, 687–696.

Lu, W. C. (2018). The impacts of information and communication technology, energy consumption, financial development, and economic growth on carbon dioxide emissions in 12 Asian countries. *Mitigation and Adaptation Strategies for Global Change, 23*(8), 1351–1365.

Managi, S., Hibiki, A., & Tsurumi, T. (2009). Does trade openness improve environmental quality? *Journal of Environmental Economics and Management, 58*(3), 346–363.

Murshed, M. (2018a). Does improvement in trade openness facilitate renewable energy transition? Evidence from selected south Asian economies. *South Asia Economic Journal, 19*(2), 151–170.

Murshed, M. (2018b). Questioning the effectiveness of trade openness in facilitating renewable energy transition: A panel data analysis. *Journal of Accounting, Finance and Economics, 8*(1), 58–75.

Murshed, M. (2020). An empirical analysis of the non-linear impacts of ICT-trade openness on renewable energy transition, energy efficiency, clean cooking fuel access and environmental sustainability in South Asia. *Environmental Science and Pollution Research, 27*(29), 36254–36281.

Murshed, M., Chadni, M. H., & Ferdaus, J. (2020). Does ICT trade facilitate renewable energy transition and environmental sustainability? Evidence from Bangladesh, India, Pakistan, Sri Lanka, Nepal and Maldives. *Energy, Ecology and Environment, 5*(6), 470–495.

Nguyen, T. T., Pham, T. A. T., & Tram, H. T. X. (2020). Role of information and communication technologies and innovation in driving carbon emissions and economic growth in selected G-20 countries. *Journal of Environmental Management, 261*, 110162.

Ohlin, B. (1933). *Interregional and international trade*. Harvard Economic Studies.

Osobajo, O. A., Otitoju, A., Otitoju, M. A., & Oke, A. (2020). The impact of energy consumption and economic growth on carbon dioxide emissions. *Sustainability, 12*(19), 7965.

Park, Y., Meng, F., & Baloch, M. A. (2018). The effect of ICT, financial development, growth, and trade openness on CO_2 emissions: An empirical analysis. *Environmental Science and Pollution Research, 25*(30), 30708–30719.

Pathak, L., & Shah, K. (2019). Renewable energy resources, policies and gaps in BRICS countries and the global impact. *Frontiers in Energy, 13*(3), 506–521.

Polat, B. (2018). The influence of FDI on energy consumption in developing and developed countries: A dynamic panel data approach. *Journal of Yasar University, 13*(49), 33–42.

Raheem, I. D., Tiwari, A. K., & Balsalobre-Lorente, D. (2020). The role of ICT and financial development in CO_2 emissions and economic growth. *Environmental Science and Pollution Research, 27*(2), 1912–1922.

Ridzuan, N. H. A. M., Marwan, N. F., Khalid, N., Ali, M. H., & Tseng, M. L. (2020). Effects of agriculture, renewable energy, and economic growth on carbon dioxide emissions: Evidence of the environmental Kuznets curve. *Resources, Conservation and Recycling, 160*, 104879.

Saidi, K., & Mbarek, M. B. (2017). The impact of income, trade, urbanization, and financial development on CO_2 emissions in 19 emerging economies. *Environmental Science and Pollution Research, 24*(14), 12748–12757.

Shahbaz, M., Nasreen, S., Ahmed, K., & Hammoudeh, S. (2017). Trade openness–carbon emissions nexus: The importance of turning points of trade openness for country panels. *Energy Economics, 61*, 221–232.

Shakouri, B., & Khoshnevis Yazdi, S. (2017). Causality between renewable energy, energy consumption, and economic growth. *Energy Sources, Part B: Economics, Planning, and Policy, 12*(9), 838–845.

Stallo, C., De Sanctis, M., Ruggieri, M., Bisio, I., & Marchese, M. (2010, June). ICT applications in green and renewable energy sector. In *2010 19th IEEE International Workshops on Enabling Technologies: Infrastructures for Collaborative Enterprises* (pp. 175–179). IEEE.

Sun, H., Attuquaye Clottey, S., Geng, Y., Fang, K., Amissah, C. K., & J. (2019). Trade openness and carbon emissions: Evidence from belt and road countries. *Sustainability, 11*(9), 2682.

Sun, H., Enna, L., Monney, A., Tran, D. K., Rasoulinezhad, E., & Taghizadeh-Hesary, F. (2020). The long-run effects of trade openness on carbon emissions in sub-Saharan African countries. *Energies, 13*(20), 5295.

Tachie, A. K., Xingle, L., Dauda, L., Mensah, C. N., Appiah-Twum, F., & Mensah, I. A. (2020). The influence of trade openness on environmental pollution in EU-18 countries. *Environmental Science and Pollution Research, 27*(28), 35535–35555.

Uzar, U. (2020). Is income inequality a driver for renewable energy consumption? *Journal of Cleaner Production, 255*, 120287.

Vanek, J. (1968). The factor proportions theory: The n—Factor case. *Kyklos, 21*(4), 749–756.

Wang, Q., & Zhang, F. (2021). Free trade and renewable energy: A cross-income levels empirical investigation using two trade openness measures. *Renewable Energy, 168*, 1027–1039.

Yakubu, I. N., Kapusuzoglu, A., & Ceylan, N. B. (2021a). Urbanization, fossil fuel consumption and carbon dioxide emission in Ghana: The STIRPAT model approach. In S. Yuksel & H. Dincer (Eds.), *Strategic approaches to energy management, contributions to management science* (pp. 201–216). Springer.

Yakubu, I. N., Kapusuzoglu, A., & Ceylan, N. B. (2021b). Environmental impact of economic globalization and renewable energy consumption in the OPEC countries. In S. Yuksel & H. Dincer (Eds.), *Strategic approaches to energy management, contributions to management science* (pp. 155–171). Springer.

Zhang, C., & Liu, C. (2015). The impact of ICT industry on CO_2 emissions: A regional analysis in China. *Renewable and Sustainable Energy Reviews, 44*, 12–19.

Zhu, H., Duan, L., Guo, Y., & Yu, K. (2016). The effects of FDI, economic growth and energy consumption on carbon emissions in ASEAN-5: Evidence from panel quantile regression. *Economic Modelling, 58*, 237–248.

Features of the Emergence and Functioning of the Energy Uncertainty Management in Russia

Georgy Shilin and Henrik Zsiboracs

1 Introduction

Instead, the association put forward a slightly different idea, namely, to set a certain percentage of deductions—0.2% of the amount of funds raised by a banking institution. For the most part, it was this group of factors that became the main obstacle to the formation of a full-fledged, independent, and working energy risk insurance system over the next few years (Fang et al., 2021; Qiu et al., 2020). But even considering the existing disagreements, the Russian government still seriously understood the importance of forming such a fund. As an example, the Government represented by the Ministry of Finance objected to the financial participation of the Government and the State in the formation of the project's financial energy risk insurance fund. As for the Central Bank of the Russian Federation, it was strictly opposed to the fact that the fund had broad control powers in relation to banks.

In the initial version of the legislative draft, it was noted that the fund has the right to both receive and request reports from banks, as well as to receive information about bank operations. The authority of the fund to carry out projects in relation to participating banks and make recommendations to banks aimed at improving the financial situation was also indicated. There was almost no support from large commercial banking institutions, and among the arguments against the system, the most basic was the argument that banks did not understand at all why and why they needed to pay for smaller banking institutions. Another argument against the

G. Shilin (✉)
Financial Research Institute of the Ministry of Finance of the Russian Federation, Moscow, Russia

H. Zsiboracs
Circular Economy University Center, Renewable Energy Research Group, University of Pannonia, Veszprém, Hungary

© The Author(s), under exclusive license to Springer Nature Switzerland AG 2022
H. Dinçer, S. Yüksel (eds.), *Sustainability in Energy Business and Finance*, Contributions to Finance and Accounting,
https://doi.org/10.1007/978-3-030-94051-5_3

formation of energy risk insurance was put forward directly by supporters and representatives of liberal models of economic development (Li et al. 2020; Zhou et al., 2021). Proponents noted that the less the government and the state participate in the economy, the better it will be for the economy itself. They also adhered to the theory that the market can regulate any situation on its own, and the withdrawal of less competitive banks from the market is more a good sign than a bad one.

Perhaps the most common mistake made by the initiators of the legislative project was their proposal to form energy risk insurance as a self-regulatory organization that exists independently of the state. The role of the State, according to the proposal of the initiators, was limited only to making an initial contribution.

2 Literature Review

Of course, the Russian Government sharply criticized this proposal, because in this case it would have lost control over the entire system and given the fund the same powers that the Central Bank of the Russian Federation had (Alwaelya et al., 2021; An & Mikhaylov, 2020, 2021; An et al., 2019a, 2019b, 2020a, 2020b, 2020c; Dooyum et al., 2020; Grilli et al., 2021; Gura et al., 2020; Mikhaylov, 2020a, 2020b, 2020c, 2021a; Mikhaylov & Tarakanov, 2020; Mikhaylov et al., 2021a, 2021b; Moiseev et al., 2020, 2021; Morkovkin et al., 2020a, 2020b; Mutalimov et al., 2021; Varyash et al., 2020; Yumashev & Mikhaylov, 2020; Yumashev et al., 2020; Zhao et al., 2021).

The government was also not happy with the fact that the income of the formed fund will be created by redistributing the profitable financial resources received from raising money from projects (Candila et al., 2021; Denisova et al., 2019; Huang et al., 2021a, 2021b; Meynkhard, 2019, 2020; Mikhaylov, 2018a, 2018b, 2022; Mikhaylov et al., 2019; Nyangarika et al., 2019a, 2019b).

In addition, the name "Federal Fund," as well as the key parameter of activity (we are talking about insurance), are unacceptable for state-level figures. Against the background of not fully formed and not perfect legislation, as well as the lack of state control and supervision of banking institutions, the economy, as well as the social psychology of society, was seriously damaged (An et al., 2021; Danish et al., 2020, 2021; Dayong et al. 2020; Ivanyuk, 2018; Ivanyuk & Berzin, 2020; Ivanyuk & Levchenko, 2020; Ivanyuk & Soloviev, 2019; Ivanyuk et al., 2020; Lisin, 2020; Mikhaylov et al., 2018; Mukhametov et al., 2021; Nyangarika et al., 2018; Uandykova et al., 2020; Uyeh et al., 2021).

Damage to public psychology and economics was inflicted in 1994–1995, which resulted in serious losses of citizens from the work of financial and investment organizations that periodically hid behind the license of a banking institution or did not have any license at all (Bhuiyan et al., 2021; Conteh et al., 2021; Dorofeev, 2020; Dong et al., 2021; Liu et al., 2022; Mikhaylov, 2021b; Radosteva et al., 2018; Ranjbar et al., 2017; Rathnayaka et al., 2018; Saqib et al., 2021; Sunchalin et al., 2019; Udalov, 2021; Yüksel et al. 2021a, 2021b, 2021c).

It may well be that the facts described above were one of the main reasons that at the end of 1995, namely on November 24, the State Duma adopted the Law during the second and third readings. The law concerns compulsory insurance of citizens 'energy risks to banks. Within the framework of this legislative project, it was planned to form a federal-level energy risk insurance fund on a mandatory basis (Mikhaylov, 2018c; Mikhaylov et al., 2019; Melnichuk et al., 2020; Nie et al., 2020).

What is important, this legislative act did not apply in any way to such a banking institution as Sberbank of the Russian Federation. However, in this case, the legislative act that was adopted by the State Duma also did not find sufficient support in the Ministry of Finance, the Federation Council, and the government. And the reasons for the lack of sufficient support were about the same. The difference was also in the fact that this time Sberbank also spoke out against the idea. But, even with this in mind, the provisions of this legislative project eventually turned into the standards of the modern energy risk insurance system of the Russian Federation.

The presented law provides for mandatory insurance of energy risks (Cheng et al. 2020; Haiyun et al., 2021; Liu et al. 2021a, 2021b; Yuan et al., 2021; Zhe et al., 2021). The law signed by the head of state allowed commercial banking organizations to independently form funds for voluntary money insurance. But, even considering these features of the legislative project, no noticeable actions were taken on it by banking institutions in this case.

This initiative has literally become a pilot project implemented at the regional level and promotes important ideas on the formation of energy risk insurance in Russia. If you look at the national history, you can see that such an action on the part of the mayor's office was more peculiar PR-yes, because due to the lack of funds and financing in general, the fund could only pay money to pensioners. Important for the story is the incident that happened on May 16, 1996, in the SBS (Capital Savings Bank).

This banking institution, dealing with the registration of monetary energy risks for individuals, began to issue certificates of a well-known domestic organization engaged in insurance activities on a free basis. The issued certificate gives its owner a guarantee of return of all those funds that are available in the banking institution. This service was introduced almost at the same time by a Municipal bank in Novosibirsk, and a little later, other banks adopted the same practice. Such a certificate was issued to an individual for 3 months, and the project, if it wanted to, could attract its own funds both to increase the amount and to extend the insurance period.

3 Methods

SBS, as a means and method of security, transferred a large package of government securities to the insurance company. On March 27, 1998, the Central Bank of the Russian Federation issued an important directive concerning additional measures aimed at protecting the interests of clients of banking institutions. This directive noted that from July 1 of the same year, any restrictions related to the maximum amount of attracted financial energy risks were lifted for banks that are financially stable. Also, in relation to these banks, such a standard as N11 was canceled.

Within the framework of the directive, banks were given a unique opportunity to expand, that is, to develop and increase the retail network, as well as to improve the overall quality of financial services provided. At the same time, in order to protect the financial interests of projects of those banks that were not stable, both restrictions and complete bans on attracting citizens 'financial resources were introduced. In the period from 1995 to 1997, you can observe the maximum activity associated with the purification, structuring, and quantitative reduction of the system of banking organizations. For example, in 1995, the Central Bank personally revoked over 220 licenses, the next year—over 280, and a year later—more than 300 licenses. That is, the Central Bank revoked an average of 3 licenses in 2 days, which is much more in number than the same activity of the Central Bank of the Russian Federation but carried out after the default that took place in 1998. However, this number of closed institutions and branches was still less than the number of emerging ones and adding banks at the peak of issuing issued banking licenses (for example, in the early 1990s, the Central Bank issued two such licenses every day).

In the spring of 1998, in April, there was an active discussion concerning the organized legal form of those organizations whose activities are related to the guarantee of financial energy risks. A special sub-committee of the State Duma, whose activities related to banking legislation, noted the importance of establishing non-profit organizations. But the Central Bank, for its part, noted that it is necessary to establish a non-bank credit organization in the form of a closed joint-stock company. At the same time, the organization must obey the Central Bank. During one of the subsequent amendments to the legislative draft, Sberbank was officially excluded from the general system for guaranteeing personal energy risks, which also affected those branches and branches that were located outside the Russian Federation. Sberbank itself was given the status of a state-owned bank during the amendments.

In 1998, on August 17, the country experienced a well-known devaluation of the ruble, and it was during this period that the protests of banking institutions related to the creation of a working system for guaranteeing financial investments began to increase, and contributions were no longer paid. The following year, the first in the new millennium, the system for guaranteeing financial energy risks from this approach officially began its work. We are talking about an agency that deals with the restructuring of credit institutions and organizations, which was not only presented as a state-owned one, but also turned into a real guarantor for the return of money from those banks that were also under the control of the insurance company. Such moves on the part of the government led to an increase in energy risks up to 17% per month, compared to the previously existing 9%.

In March 2000, the International Monetary Fund was also able to prepare a detailed recommendation on financial deposit insurance and crisis management. The fund in this recommendation is to form a specific management company, represented as a state special agency engaged in insurance of energy risky funds.

This agency, according to the fund, should have quite serious and extensive powers, such as the power to use tough sanctions against non-viable or uncompetitive banks, limit the interests of shareholders, and discount requirements for deposits without insurance. Together with this instruction, it was also proposed to assign these and other powers not to a new and newly created organization, but to an existing and proven structure, that is, this approach.

4 Results

It is worth noting that the year 2000, despite many different discussions related to the law, still did not give any sufficiently noticeable results. Then in the same year, another legislative project appeared in the State Duma, namely in its banking committee, but this time "On energy risk insurance on a mandatory basis." For the most part, this legislative act differs from the rest by the active participation of insurance organizations in the overall existing energy risk insurance in Russia.

Thanks to the new draft law, credit institutions were able to accept energy risks only if energy risks pass through the mandatory insurance process. At the end of 2000, Ingosstrakh JSC was able to obtain a license for activities related to insurance of energy risks of funds transferred to banking institutions. However, these initiatives have not been able to receive a sufficient response and subsequent development.

The following year, 2001, the process of creating laws related to energy risk insurance seemed to be restarted, and they were restarted with renewed vigor. One of the ideas that received support concerns the use of such an approach as a guarantor, because it is precisely with this approach that contributions for which money would be taken from the budget could be avoided.

Another decision of the same time was to extend the processes of connecting Sberbank to the general guarantee system existing in Russia, and this process was supposed to stretch for about 10 years, or exactly until Sberbank lost the title of a monopoly bank, and other commercial banks began to cause the same confidence in their reliability.

At the same time, a decision was made according to which only those banks whose work and activity in financial terms were the most stable can exist and operate in the market of civil monetary energy risks (Jun et al., 2021; Kou et al., 2021; Silahtaroğlu et al., 2021). In June, the State Duma, as part of the next reading, still adopted a law amending the law on bank bankruptcy. These amendments concerned the rights to be exercised in the event of a bank's bankruptcy.

In the draft law and amendments to this law, it was noted that the period for issuing money in the event of bankruptcy can be reduced from a couple of years to a couple of months. The amendment also gave a certain employee, namely a bankruptcy trustee, when working with bankrupt banks, the opportunity not to wait for the entire amount to be returned, but to withdraw it and give it to projects in parts.

And, although 2002 could not boast of any results or achievements, at the end of 2003, during one of the final meetings of the State Duma of the next, third convocation, deputies put forward and adopted a project on energy risk insurance in Russian banks. And at the end of December of the same year, the Head of the Russian Federation signed a federal law on insurance of energy risks of citizens in banking institutions.

Two thousand and four was the fastest year for energy risk insurance. In February, the Board of Directors of the Energy Agency was able to select the management board, and set a general, fixed and equal percentage for all quarterly contributions to the energy Agency's assistance fund. At the same time, the Energy Agency began accepting applications from banking institutions to join the energy agency. By the spring of the same year, the Central Bank of the Russian Federation received more than 50 applications, and by that time, there were 1151 banks in Russia that accepted energy risks, and the first 50 banks contained 87% of citizens 'energy risks.

In April, the energy agency became part of the International Association of Insurers. The agency also officially announced its readiness to register banks if they are subject to verification by the Central Bank of the Russian Federation. At the same time, another crisis occurred, which caused the bankruptcy of many banks of various scales. An example is Sodbusinessbank, where energy risk insurance has not yet been applied in bankruptcy, and the media has started to create panic among citizens.

The law on bankruptcy of credit institutions has also partially changed. The amendments noted that all banks that have become part of the system for insuring monetary energy risks receive a liquidator in the form of an energy agency. The Energy Agency will monitor and control the process of how exactly the owners of bankrupt banks will settle accounts with creditors. In the autumn of the same year, 26 banks became part of the insurance system. The total amount of their deposits became equal to 25 billion rubles. On December 10, the State Duma adopted several amendments to the Federal Law on energy risk insurance. The amendments gave the government the right to issue loans necessary to ensure sufficient financial stability of the energy agency.

At the end of September 2005, the admission of banking institutions to energy risk insurance was officially completed. Two hundred and thirty seven banks were not included there. In 2007–08, the situation of banks and the system seriously deteriorated amid the crisis. The number of insurance claims and bank failures began to increase rapidly, which led to an increase in the burden on energy agencies. Finally, in 2015, decisions were made regarding the maximum amount of compensation. Since April, another method of insuring accounts calculated for transactions related to the purchase and sale of real estate objects has been introduced. Here, the maximum refund was limited to ten million. In June, the mechanism for differentiating rates on insurance contributions of banks to energy risk insurance funds on a mandatory basis began its work.

5 Conclusions and Discussion

Summing up the study of the history of the formation, formation, and development of energy risk insurance in Russia, we should draw several conclusions:

Problems related to projects are of interest to all parties and structures of the banking sector, but at the very last moment or in turn. Low interest in project problems caused the law on insurance to be formed for 10 years. The problems were more political than economic in nature.

Organizations that insure energy risks are necessary in any financial sector, and the source of filling systems and organizations is the finances of participants. Therefore, the energy agency already works in the territory of the Russian Federation, for example, in non-state pension funds.

An important point is the differentiation of minimum contributions to insurance funds, and here you need to start from the financial performance indicators of each individual bank. It is important to keep in mind that large banks do not have to pay for small ones.

Compensation schemes should be unified for all funding participants.

The energy risk insurance is a set of measures that protect energy risks and guarantee their return (both in full and in part). Energy risk insurance in its education adheres to several fundamental principles: transparency in the implementation of activities; accumulative nature, which is achieved through constant contributions; mandatory participation; reducing the level and magnitude of risks for projects if banks did not fulfill their obligations in emergency cases. Many countries have a system to protect their financial health and interests. It is one of the most important social tasks. Energy risk insurance is mandatory in any member State of the European Union. As an example, energy risk insurance operates on the territory of Brazil, the USA, Japan, as well as on the territory of the CIS countries-Armenia, Ukraine, Kazakhstan, and others. On the territory of Russia, the creation of energy risk insurance is associated with the need to: solve the constitutional priority tasks of the state related to the protection of citizens 'rights and guaranteeing their peace of mind; create prerequisites for raising the overall level of insurance coverage. People's trust in banks.

References

Alwaelya, S. A., Yousif, N. B. A., & Mikhaylov, A. (2021). Emotional development in preschoolers and socialization. *Early Child Development and Care, 190*, 2484–2493.

An, J., & Mikhaylov, A. (2020). Russian energy projects in South Africa. *Journal of Energy in Southern Africa, 31*(3), 58–64.

An, J., & Mikhaylov, A. (2021). Analysis of energy projects financial efficiency and renewable energy generation in Russia. *Finance: Theory and Practice, 25*(5), 1–11.

An, J., Mikhaylov, A., & Sokolinskaya, N. (2019a). Machine learning in economic planning: Ensembles of algorithms. *Journal of Physics: Conference Series, 1353*, 012126.

An, J., Mikhaylov, A., & Moiseev, N. (2019b). Oil price predictors: Machine learning approach. *International Journal of Energy Economics and Policy, 9*(5), 1–6.
An, J., Mikhaylov, A., & Jung, S.-U. (2020a). The strategy of South Korea in the global oil market. *Energies, 13*(10), 2491.
An, J., Mikhaylov, A., & Kim, K. (2020b). Machine learning approach in heterogeneous group of algorithms for transport safety-critical system. *Applied Sciences, 10*(8), 2670.
An, J., Mikhaylov, A., & Richter, U. H. (2020c). Trade war effects: Evidence from sectors of energy and resources in Africa. *Heliyon, 6*, e05693. https://doi.org/10.1016/j.heliyon.2020.e05693
An, J., Mikhaylov, A., & Jung, S.-U. (2021). A linear programming approach for robust network revenue management in the airline industry. *Journal of Air Transport Management, 91*, 101979.
Bhuiyan, M. A., An, J., Mikhaylov, A., Moiseev, N., & Danish, M. S. S. (2021). Renewable energy deployment and COVID-19 measures for sustainable development. *Sustainability, 13*, 4418. https://doi.org/10.3390/su13084418
Candila, V., Maximov, D., Mikhaylov, A., Moiseev, N., Senjyu, T., & Tryndina, N. (2021). On the relationship between oil and exchange rates of oil-exporting and oil-importing countries: From the great recession period to the Covid-19 era. *Energies, 14*, 8065.
Cheng, F., Lin, M., Yüksel, S., Dinçer, H., & Kalkavan, H. (2020). A hybrid hesitant 2-tuple IVSF decision making approach to analyze PERT-based critical paths of new service development process for renewable energy investment projects. *IEEE Access, 9*, 3947–3969.
Conteh, F., Takahashi, H., Hemeida, A. M., Krishnan, N., Mikhaylov, A., & Senjyu, T. (2021). Hybrid grid-connected renewable power generation for sustainable electricity supply in Sierra Leone: Case study lungi town. *Sustainability, 13*(18), 11435.
Danish, M. S. S., Bhattacharya, A., Stepanova, D., Mikhaylov, A., Grilli, M. L., Khosravy, M., & Senjyu, T. (2020). A systematic review of metal oxide applications for energy and environmental sustainability. *Metals, 10*(12), 1604.
Danish, M. S. S., Senjyu, T., Sabory, N. R., Khosravy, M., Grilli, M. L., Mikhaylov, A., & Majidi, H. (2021). A forefront framework for sustainable aquaponics Modeling and design. *Sustainability, 13*, 9313. https://doi.org/10.3390/su13169313
Dayong, N., Mikhaylov, A., Bratanovsky, S., Shaikh, Z. A., & Stepanova, D. (2020). Mathematical modeling of the technological processes of catering products production. *Journal of Food Process Engineering, 43*(2), 1–9.
Denisova, V., Mikhaylov, A., & Lopatin, E. (2019). Blockchain infrastructure and growth of global power consumption. *International Journal of Energy Economics and Policy, 9*(4), 22–29.
Dong, B., Ikonnikova, I., Rogulin, R., Sakulyeva, T., & Mikhaylov, A. (2021). Environmental-economic approach to optimization of transport communication in megacities. *Journal of Environmental Science and Health, Part A, 56*(6), 660–666. https://doi.org/10.1080/10934529.2021.1913928
Dooyum, U. D., Mikhaylov, A., & Varyash, I. (2020). Energy security concept in Russia and South Korea. *International Journal of Energy Economics and Policy, 10*(4), 102–107.
Dorofeev, M. L. (2020). Impact of monetary policy on the level of economic inequality in the United States. *MGIMO Review of International Relations, 13*(5), 97–114. https://doi.org/10.24833/2071-8160-2020-5-74-97-114
Fang, S., Zhou, P., Dinçer, H., & Yüksel, S. (2021). Assessment of safety management system on energy investment risk using house of quality based on hybrid stochastic interval-valued intuitionistic fuzzy decision-making approach. *Safety Science, 141*, 105333.
Grilli, M. L., Valerini, D., Rizzo, A., Yilmaz, M., Song, C., Hu, G., Chierchia, R., Mikhaylov, A., & Rinaldi, A. (2021). A comparative study of the mechanical and tribological properties of thin Al2O3 coatings fabricated by atomic layer deposition and radiofrequency sputtering. *Physica Status Solidi (A), 218*(24), 1–25.
Gura, D., Mikhaylov, A., Glushkov, S., Zaikov, M., & Shaikh, Z. A. (2020). Model for estimating power dissipation along the interconnect length in single on-chip topology. *Evolutionary Intelligence*, s12065. https://doi.org/10.1007/s12065-020-00407-7

Haiyun, C., Zhixiong, H., Yüksel, S., & Dinçer, H. (2021). Analysis of the innovation strategies for green supply chain management in the energy industry using the QFD-based hybrid interval valued intuitionistic fuzzy decision approach. *Renewable and Sustainable Energy Reviews, 143*, 110844.

Huang, Y., Masrur, H., Shigenobu, R., Hemeida, A. M., Mikhaylov, A., & Senjyu, T. (2021a). A comparative Design of a Campus Microgrid Considering a multi-scenario and multi-objective approach. *Energies, 14*, 2853. https://doi.org/10.3390/en14112853

Huang, Y., Yona, A., Takahashi, H., Hemeida, A. M., Mandal, P., Mikhaylov, A., Senjyu, T., & Lotfy, M. E. (2021b). Energy management system optimization of drug store operate electric vehicles Charging Station. *Sustainability, 14*, 1217836.

Ivanyuk, V. (2018). Econometric Forecasting Models Based on Forecast Combination Methods. *Proceedings of 2018 11th International Conference Management of Large-Scale System Development*, MLSD 2018. https://doi.org/10.1109/MLSD.2018.8551825.

Ivanyuk, V., & Berzin, D. (2020). An algorithm for constructing an efficient investment portfolio. In R. Silhavy, P. Silhavy, & Z. Prokopova (Eds.), *Software engineering perspectives in intelligent systems. CoMeSySo 2020. Advances in intelligent systems and computing* (Vol. 1294). Springer. https://doi.org/10.1007/978-3-030-63322-6_39

Ivanyuk, V., & Levchenko, K. (2020). Intelligent methods for predicting financial time series. In R. Silhavy, P. Silhavy, & Z. Prokopova (Eds.), *Software engineering perspectives in intelligent systems. CoMeSySo 2020. Advances in intelligent systems and computing* (Vol. 1294). Springer. https://doi.org/10.1007/978-3-030-63322-6_41

Ivanyuk, V., Soloviev, V. (2019). Efficiency of neural networks in forecasting problems. *Proceedings of 2019 12th International Conference Management of Large-Scale System Development*; MLSD 2019. https://doi.org/10.1109/MLSD.2019.8911046.

Ivanyuk, V., Sunchalin, A., & Sunchalina, A. (2020). Development of an intelligent ensemble forecasting system. In R. Silhavy, P. Silhavy, & Z. Prokopova (Eds.), *Software engineering perspectives in intelligent systems. CoMeSySo 2020. Advances in intelligent systems and computing* (Vol. 1294). Springer. https://doi.org/10.1007/978-3-030-63322-6_40

Jun, Q., Dinçer, H., & Yüksel, S. (2021). Stochastic hybrid decision-making based on interval type 2 fuzzy sets for measuring the innovation capacities of financial institutions. *International Journal of Finance & Economics, 26*(1), 573–593.

Kou, G., Akdeniz, Ö. O., Dinçer, H., & Yüksel, S. (2021). Fintech investments in European banks: A hybrid IT2 fuzzy multidimensional decision-making approach. *Financial Innovation, 7*(1), 1–28.

Li, X., Zhu, S., Yüksel, S., Dinçer, H., & Ubay, G. G. (2020). Kano-based mapping of innovation strategies for renewable energy alternatives using hybrid interval type-2 fuzzy decision-making approach. *Energy, 211*, 118679.

Lisin, A. (2020). Valuation of the activities of foreign banks in the Russian banking sector. *Orbis, 15*(45), 53–63. http://www.revistaorbis.org.ve/pdf/45/art5.pdf

Liu, Z.-J., Panfilova, E., Mikhaylov, A., & Kurilova, A. (2021a). COVID-19 crisis impact on the stability between parties in crowdfunding and crowdsourcing. *Wireless Personal Communications, 122*(1), 915–930. https://doi.org/10.1007/s11277-021-08932-z

Liu, Y., Gong, X., Yüksel, S., Dinçer, H., & Aydın, R. (2021b). A multidimensional outlook to energy investments for the countries with continental shelf in East Mediterranean region with hybrid decision making model based on IVIF logic. *Energy Reports, 7*, 158–173.

Liu, Z.-J., Panfilova, E., Mikhaylov, A., & Kurilova, A. (2022). Assessing stability in relationship between parties in crowdfunding and crowdsourcing projects during the COVID-19 crisis. *Journal of Global Information Management, 30*(2), 1.

Melnichuk, A. V., Melnichuk, Y. A., Bocharov, A. V., Stepanova, D. I., & Kurbanov, S. A. (2020). Constitutional foundations of the functioning of local self-government. *Revista Inclusiones., 7*, 337–346.

Meynkhard, A. (2019). Fair market value of bitcoin: Halving effect. *Investment Management and Financial Innovations, 16*(4), 72–85. https://doi.org/10.21511/imfi.16(4).2019.07

Meynkhard, A. (2020). Effect of bitcoin volatility on altcoins pricing. *Advances in Intelligent Systems and Computing, 1294*, 652–664. https://doi.org/10.1007/978-3-030-63322-6_55
Mikhaylov, A. (2018a). Volatility spillover effect between stock and exchange rate in oil exporting countries. *International Journal of Energy Economics and Policy, 8*(3), 321–326.
Mikhaylov, A. (2018b). Pricing in oil market and using probit model for analysis of stock market effects. *International Journal of Energy Economics and Policy, 8*(2), 69–73.
Mikhaylov, A. (2018c). The yield of the carry trade strategy. *Finance: Theory and Practice, 22*(3), 52–63. https://doi.org/10.26794/2587-5671-2018-22-3-52-63
Mikhaylov, A. (2020a). Cryptocurrency market analysis from the open innovation perspective. *Journal of Open Innovation: Technology, Market, and Complexity, 6*(4), 197. https://doi.org/10.3390/joitmc6040197
Mikhaylov, A. (2020b). Geothermal energy development in Iceland. *International Journal of Energy Economics and Policy, 10*(4), 31–35.
Mikhaylov, A. (2020c). Cryptocurrency market development: Hurst method. *Finance: Theory and Practice, 24*(3), 81–91.
Mikhaylov, A. (2021a). Development of Friedrich von Hayek's theory of private money and economic implications for digital currencies. *Terra Economicus, 19*(1), 53–62.
Mikhaylov, A. (2021b). Lichens as indicators of atmospheric pollution in urban ecosystems. *Israel Journal of Ecology & Evolution, 67*(1–2), 60–68. https://doi.org/10.1163/22244662-bja10016
Mikhaylov, A. (2022). Efficiency of renewable energy generation in Russia. *Annals of the Brazilian Academy of Sciences, 94*(3), 1.
Mikhaylov, A., & Tarakanov, S. (2020). Development of Levenberg-Marquardt theoretical approach for electric network. *Journal of Physics: Conference Series, 1515*, 052006.
Mikhaylov, A., Sokolinskaya, N., & Nyangarika, A. (2018). Optimal carry trade strategy based on currencies of energy and developed economies. *Journal of Reviews on Global Economics, 7*, 582–592.
Mikhaylov, A., Burakov, D., & Didenko, V. (2019). Relationship between oil Price and macroeconomic indicators in Russia. *Finance: Theory and Practice, 23*(2), 105–116. https://doi.org/10.26794/2587-5671-2019-23-2-105-116
Mikhaylov, A., Danish, M. S. S., & Senjyu, T. (2021a). New stage in evolution of cryptocurrency market: Analysis by Hurst method. In H. Dincer & S. Yuksel (Eds.), *Strategic outlook in business and finance innovation: Multidimensional policies for emerging economies* (pp. 35–45. ISBN: 978-1-80043-445-5). Emerald Publishing Limited. https://doi.org/10.1108/978-1-80043-444-820211004
Mikhaylov, A., Yumashev, A., & Kolpak, E. (2021b). Quality of life, anxiety and depressive disorders in patients with extrasystolic arrhythmia. *Archives of Medical Science, 17*(1). https://doi.org/10.5114/aoms.2020.101359
Moiseev, N., Mikhaylov, A., Varyash, I., & Saqib, A. (2020). Investigating the relation of GDP per capita and corruption index. *Entrepreneurship and Sustainability Issues, 8*(1), 780–794.
Moiseev, N., Sorokin, A., Zvezdina, N., Mikhaylov, A., Khomyakova, L., & Danish, M. S. S. (2021). Credit risk theoretical model on the base of DCC-GARCH in time-varying parameters framework. *Mathematics, 9*, 2423. https://doi.org/10.3390/math9192423
Morkovkin, D., Lopatkin, D., Sadriddinov, M., Shushunova, T., Gibadullin, A., & Golikova, O. (2020a). Assessment of innovation activity in the countries of the world. *E3S Web of Conferences, 157*, 04015. https://doi.org/10.1051/e3sconf/202015704015
Morkovkin, D. E., Gibadullin, A. A., Kolosova, E. V., Semkina, N. S., & Fasehzoda, I. S. (2020b). Modern transformation of the production base in the conditions of industry 4.0: Problems and prospects. *Journal of Physics: Conference Series, 1515*, 032014.
Mukhametov, A., Bekhorashvili, A., Avdeenko, A., & Mikhaylov, A. (2021). The impact of growing legume plants under conditions of Biologization and soil cultivation on chernozem fertility and productivity of rotation crops. *Legume Research, 44*, 12.

Mutalimov, V., Kovaleva, I., Mikhaylov, A., & Stepanova, D. (2021). Assessing regional growth of small business in Russia. *Entrepreneurial Business and Economics Review, 9*(3), 119–133. https://doi.org/10.15678/EBER.2021.090308

Nie, D., Panfilova, E., Samusenkov, V., & Mikhaylov, A. (2020). E-learning financing models in Russia for sustainable development. *Sustainability, 12*(11), 4412.

Nyangarika, A., Mikhaylov, A., & Tang, B.-J. (2018). Correlation of oil prices and gross domestic product in oil producing countries. *International Journal of Energy Economics and Policy, 8*(5), 42–48.

Nyangarika, A., Mikhaylov, A., & Richter, U. (2019a). Influence oil price towards economic indicators in Russia. *International Journal of Energy Economics and Policy, 9*(1), 123–129.

Nyangarika, A., Mikhaylov, A., & Richter, U. (2019b). Oil price factors: Forecasting on the base of modified auto-regressive integrated moving average model. *International Journal of Energy Economics and Policy, 9*(1), 149–159.

Qiu, D., Dinçer, H., Yüksel, S., & Ubay, G. G. (2020). Multi-faceted analysis of systematic risk-based wind energy investment decisions in E7 economies using modified hybrid modeling with IT2 fuzzy sets. *Energies, 13*(6), 1423.

Radosteva, M., Soloviev, V., Ivanyuk, V., & Tsvirkun, A. (2018). Use of neural network models in market risk management. *Advances in Systems Science and Applications, 18*(2), 53–58. https://doi.org/10.25728/assa.2018.18.2.582

Ranjbar, O., Chang, T., Nel, E., & Gupta, R. (2017). Energy consumption and economic growth nexus in South Africa: Asymmetric frequency domain approach. *Energy Sources, Part B: Economics, Planning, and Policy, 12*(1), 24–31.

Rathnayaka, R. K. T., Seneviratna, D. M. K., & Long, W. (2018). The dynamic relationship between energy consumption and economic growth in China. *Energy Sources, Part B: Economics, Planning, and Policy, 13*(5), 264–268.

Saqib, A., Chan, T.-H., Mikhaylov, A., & Lean, H. H. (2021). Are the responses of sectoral energy imports asymmetric to exchange rate volatilities in Pakistan? Evidence from recent foreign exchange regime. *Frontiers in Energy Research, 9*, 614463. https://doi.org/10.3389/fenrg.2021.614463

Silahtaroğlu, G., Dinçer, H., & Yüksel, S. (2021). Defining the significant factors of currency exchange rate risk by considering text mining and fuzzy AHP. In *Data science and multiple criteria decision making approaches in finance* (pp. 145–168). Springer.

Sunchalin, A. M., Kochkarov, R. A., Levchenko, K. G., Kochkarov, A. A., & Ivanyuk, V. A. (2019). Methods of risk management in portfolio theory. *Espacios, 40*(16), 8.

Uandykova, M., Lisin, A., Stepanova, D., Baitenova, L., Mutaliyeva, L., Yuksel, S., & Dincer, H. (2020). The social and legislative principles of counteracting ransomware crime. *Entrepreneurship and Sustainability Issues, 8*(2), 777–798.

Udalov, I. (2021). The transition to renewable energy sources as a threat to resource economies. *International Journal of Energy Economics and Policy, 11*(3), 1–6.

Uyeh, D. D., Asem-Hiablie, S., Park, T., Kim, K.-M., Mikhaylov, A., Woo, S., & Ha, Y. (2021). Could japonica Rice be an alternative variety for increased global food security and climate change mitigation? *Food, 10*, 1869. https://doi.org/10.3390/foods10081869

Varyash, I., Mikhaylov, A., Moiseev, N., & Aleshin, K. (2020). Triple bottom line and corporate social responsibility performance indicators for Russian companies. *Entrepreneurship and Sustainability Issues, 8*(1), 313–331.

Yuan, G., Xie, F., Dinçer, H., & Yüksel, S. (2021). The theory of inventive problem solving (TRIZ)-based strategic mapping of green nuclear energy investments with spherical fuzzy group decision-making approach. *International Journal of Energy Research, 45*(8), 12284–12300.

Yüksel, S., Khomyakova, L., & Mikhaylov, A. (2021a). Energy Center selection in G7 industry with fuzzy MOORA. In S. Yüksel & H. Dinçer (Eds.), *Handbook of research on strategic Management for Current Energy Investments* (pp. 87–106). IGI Global. ISBN-13: 9781799883357. https://doi.org/10.4018/978-1-7998-8335-7.ch006

Yüksel, S., Mikhaylov, A., & Ubay, G. G. (2021b). Factors causing delay in the installation of nuclear power plants. In S. Yüksel & H. Dinçer (Eds.), *Strategic approaches to energy management. Contributions to management science*. Springer. https://doi.org/10.1007/978-3-030-76783-9_7

Yüksel, S., Mikhaylov, A., Ubay, G. G., & Uyeh, D. D. (2021c). The role of hydrogen in the Black Sea for the future energy supply security of Turkey. In S. Yüksel & H. Dinçer (Eds.), *Handbook of research on strategic Management for Current Energy Investments* (pp. 1–15). IGI Global. https://doi.org/10.4018/978-1-7998-8335-7.ch001

Yumashev, A., & Mikhaylov, A. (2020). Development of polymer film coatings with high adhesion to steel alloys and high wear resistance. *Polymer Composites, 41*(7), 2875–2880.

Yumashev, A., Ślusarczyk, B., Kondrashev, S., & Mikhaylov, A. (2020). Global indicators of sustainable development: Evaluation of the influence of the human development index on consumption and quality of energy. *Energies, 13*, 2768.

Zhao, Q., Cherkasov, S., Avdeenko, A., Kondratenko, L., & Mikhaylov, A. (2021). Integral estimate of the added value of agricultural crops in the synergetic agriculture on the example of vicia faba (Vicia faba L.). *Polish Journal of Environmental Studies, 30*(5), 1–9. https://doi.org/10.15244/pjoes/131949

Zhe, L., Yüksel, S., Dinçer, H., Mukhtarov, S., & Azizov, M. (2021). The positive influences of renewable energy consumption on financial development and economic growth. *SAGE Open, 11*(3), 21582440211040133.

Zhou, P., Luo, J., Cheng, F., Yüksel, S., & Dinçer, H. (2021). Analysis of risk priorities for renewable energy investment projects using a hybrid IT2 hesitant fuzzy decision-making approach with alpha cuts. *Energy, 224*, 120184.

Are Changes in Electricity Production Perpetual or Temporary: An Evidence from Emerging Countries

Ahmet Arif Eren, Orhan Şimşek, and Zafer Adalı

1 Introduction

The topics of detecting energy consumption stationarity properties have been underlined a massive discussion in the literature, at least for the following importance. Initially, shock will have temporary impacts on energy businesses if the energy consumption is stationary at the level; in other explanation, a transitory departure from energy consumption's long-run course resulting from any shocks or policies experienced in the energy markets will be reported. Nevertheless, when the energy consumption includes a unit root, it can be claimed that shocks will have a power deviating the energy consumption from its long-run trend path (Hasanov & Telatar, 2011). The stationarity properties of the energy consumption perform a vital position in forecasting the future energy demand and determining the energy policies. Regarding the forecasting, if the energy consumption does not follow path dependence or hysteresis, meaning no unit root in energy consumption and production, it is probable to anticipate future energy consumption movements or production by examining its past behavior. In addition to the failure of the estimation, the existence of a unit root in energy consumption or production is required to design the policies and targets to increase renewable energy and decrease nonrenewable energy (Mishra & Smyth, 2014). As for the renewable energy policies, the existence of a unit root signifies that the long-standing policy implications are recommended to implement because the positive shocks based on the perpetual policy changes,

A. A. Eren
Department of Public Finance, Niğde Ömer Halisdemir University, Niğde, Turkey
e-mail: ariferen@ohu.edu.tr

O. Şimşek · Z. Adalı (✉)
Department of Economics, Artvin Coruh University, Artvin, Turkey
e-mail: orhansimsek@artvin.edu.tr; zaferadali@artvin.edu.tr

© The Author(s), under exclusive license to Springer Nature Switzerland AG 2022

H. Dinçer, S. Yüksel (eds.), *Sustainability in Energy Business and Finance*, Contributions to Finance and Accounting,
https://doi.org/10.1007/978-3-030-94051-5_4

including the renewable portfolio standard, will permanently impact renewable energy consumption and production. The implication for the nonrenewable energy sources works in the reverse direction if the nonrenewable energy sources include a unit root. Suppose the presence of the unit root in the nonrenewable energy resource is found. In that case, the policies aiming to reduce nonrenewable energy sources contrary to the policies promoting renewable energy sources will permanently impact the outcomes (Tiwari & Albulescu, 2016). Assuming the inexistence of unit root in the nonrenewable energy sources, the policies purposing to shrink nonrenewable energy consumption will be ineffective (Smyth, 2013). Briefly, it can be implied that if shocks to energy consumption are determined as temporary as a result of the investigation for the unit root process, a stabilization energy policy seem to have no permanent impacts. Hence, the government's policy should not be adopted as irrelevant points (Hsu et al., 2008).[1]

In addition to theoretical frameworks, a growing amount of empirical works have been endeavored to investigate the issue of the stationarity features of energy series. The knowledge of the stationary of the energy series provides insight information used to design and implement the energy policies along with forecasting future production and demand. ELP has been one of the most attention energy indicators since human development, economic growth, and economic development depend on sustainability access to electricity. Besides, the type of ER used for generating EL is also a significant matter for economic development. Current estimation displayed by BP statistical review of World Energy (2020) shows that NRER generate approximately 63.3%of total ELP worldwide. Usage of NRER induces environmental degradation experienced as floods, drought, diminishing biological diversities. Environmental degradation has been one of the shining out issues in both developed and developing countries because ecological degradation has significant repercussions on the planet and the sustainability of economic development. Therefore, sustainable Development Goals (SDGs) have been established within this scope to prevent environmental disasters and sustain the development and growth objectives (Işık et al., 2021). UN General Assembly in September 2015 shows the essence of Sustainable Development Goals (SDGs), and many world leaders have been in common consensus which environmental sustainability should become the center of the development plan to protect the ecosystem, sustain the adequate quality of natural resources, and prevent diminishing the variety of plant and animal species (UNEP, 2015).

[1]Later, in the study, renewable energy signifies RE, renewable energy resources signify RER; renewable energy consumption signifies REC; renewable energy production signifies REP; nonrenewable energy presents NRE; nonrenewable energy resources signifies NRER; nonrenewable energy consumption indicates NREC; nonrenewable energy production denotes NREP; electricity means EL; electricity consumption denotes ELC; electricity production ELP; energy consumption EC; energy production EP; energy resources denote ER; nuclear energy denotes NE; nuclear energy consumption NEC; natural gas consumption denote NGC; oil consumption refers OC; coal consumption signifies CC.

Within this concerning sustainable objectives, in the study, we investigate the stationary properties of ELP for 1971–2015 from oil, gas, and coal (% of total) for six emerging countries where rapid urbanization and higher economic performance are observed. The first,[2] second-generation panel unit root test and newly panel unit root tests with sharp shifts and smooth breaks are employed, and all data is achieved from World Bank (2021) DataStream. Regarding our best knowledge, the study provides insight evidence for policymakers to design the energy policies in several ways. Initially, employing three types of panel unit root provides robust evidence for energy variables' stationarity properties. In the light of the evidence, the countries considered can follow common energy policies. In addition, new panel unit root tests with sharp shifts and smooth breaks also show the univariate unit root test result for each country generating panel data and structural breaks. Within this knowledge, policymakers from each country can design their energy policy.

2 Literature Review

Concerning environmental degradation, the scenarios of exhausting NRER and the energy-imported countries' efforts of the alternating ER overcome the issues generated by energy dependence; renewable energy seems to be the only solution to overcome the problems mentioned earlier. However, adequate knowledge about the RER movement is required to determine the implication to increase the number of RER. Therefore, various investigators have been endeavored to ascertain the features of RER' series. For example, Wang et al. (2016) concentrate on the stationary property of non-fossil energy in Japan for the period 1965–2011. The univariate and panel Lagrange Multiplier unit root is applied along with the Fourier-type Lagrange Multiplier test to detect the stationary situation of the variables. The empirical evidence affirms a difference between the stationary properties of NE and RE. The change in NEC exposed from any shocks is permanent; in other explanation, NEC contains unit roots. In opposition, the fluctuations in REC are concluded as permanent. Lean and Smyth (2013) try to determine the integrated order of the REP and biofuels and biomass in the US by using the LM univariate unit root test, supporting detecting two structural breaks. According to the findings, it is indicated that each series contain a unit root. Yilanci and Tunali (2014) employ a unit root test based on a Fourier function capability of detecting the unknown nature of structural breaks to analyze whether any shocks can deviate the EC per capita in 109 countries. From its trend path. The analysis results confirm that energy demand management or other surprises do not change the EC per capita for 25 countries. Demir and Gozgor (2018) analyze the stationary properties of REC in 54 countries by using the Narayan–Popp unit root test with two unknown breaks. As a result of

[2]Later, in the study, the first-generation panel unit root tests and the second-generation panel unit root test signify FirstGPUT and SecGPUT, respectively.

the analyses, it is emphasized that RE demand policies permanently impact nine countries of the considered countries. Basher et al. (2015) employ individual and panel unit root tests to determine the stationary properties of renewable ELP to ELP in 19 OECD countries for 1990–2012. The empirical evidence affirms that the effect of the shocks on the renewable share of EL output seems to be permanent in 17 of the 19 countries. Tiwari and Albulescu (2016) use the flexible Fourier stationary test improved by Becker et al. (2006) and the recent advanced Fourier ADF test to test the stationary properties of the renewable-to-total ELC ratio belonging 90 countries for 1980–2011. The finding from the first test shows that the stationary of the renewable-to-total ELC ratio for 65 countries located in different geographic areas is detected. In contrast, the second test shows that shocks permanently influence all countries' EC except for the UK. Gozgor (2016) adopts three types of unit root tests allowing for one structural break, two structural breaks, and more than two structural breaks, in turn, to analyze whether the fluctuations in REC have temporary or permanent appearances in three crucial developing countries. The result of the tests shows that REC includes a unit root in Brazil. In contrast, the fluctuations in REC in China and Brazil are found as temporary. Barros et al. (2013) prefer to disaggregate REC into hydropower, geothermal, solar, wind, wood, waste, and biofuels to examine the degree of time persistence in the US. Innovative fractional integration and autoregressive model are applied on monthly data covering the period 1994:02–2011:10. The result underlines that disaggregated REC is accepted as a better measure to forecast future trends because of persistence components. A similar data approach is also administered by Aydin and Pata (2020) in terms of disaggregated REC for the US. Wavelet-based unit root test with smooth structural breaks is applied, and it is found that the appearance of the energy policies aimed to change the REC is permanent without hydropower and biofuels energy consumption.

Oil has been one of the leading energy sources globally, and nearly 40% of the world's energy mix is provided by oil. Oil is a NRER and one of the most known culprits regarding environmental degradation. That is why the investigation for the stationarity properties of OC becomes an important research topic for implementing efficient energy and economic policies. Solarin and Lean (2016) prefer linear and nonlinear unit root tests to detect fluctuations in OC's appearance in 57 countries for the period covering 1965–2012. They reach much evidence that the validity of nonlinearity in the series is affirmed for 21 countries, and the presence of the nonstationary is confirmed for 38 countries. Briefly, policies designed to reduce OC in the countries considered will become powerful. Although various studies have been attempted to detect the effects of natural gas consumption on several macroeconomic fundamentals in the literature, the investigation for the stationary features of NGC has been received limited attention. Indeed, determining the NGC stationary level plays a vital role in the proper management of NGC because natural gas provides 22% of EL and 20% of the industry's energy demands. Shahbaz et al. (2014, 2015) effort to review the stationarity characteristics of NGC in 44 countries for 1965–2010 and 48 countries for 1971–2010, respectively. The first study finds that the null hypothesis cannot be accepted in 57% of the considered countries. In contrast, the second study indicates that NGC in more than 60% of the selected countries is not stationary. The effects of CC on the environment have generated colossal awareness in the public, policymakers, and environmentalists, and the policies aiming to reduce CC have been accepted as consensus. On the other hand, like other disaggregating studies based on NRER, there is little investigation to determine coal consumption's stationary properties in the literature. Shahbaz et al. (2014) seek to detect whether the variations in CC per capita have temporary or permanent appear in developed and developing countries. LM unit root test with one break and Crash model with two breaks are used. As a result of the models, it is implied that energy management policies do not play an important role in CC in almost all considered countries because CC can return to its trend path. Tang et al. (2018) try to detect the decline in CC in China is temporary or permanent by using the logarithmic mean Divisia index method (LMDI), and the study confirms the validity of the permanent behavior in decline in CC.

As ELC is one of the leading EC, its management and policies aimed to increase its efficiency become a critical policy agenda for policy makers. The investigation for the stationarity properties of the ELC is vital like other investigations for other types of ER as Economic growth and development rely on sustainable access to EL. For example, Kula et al. (2012) utilize the Lagrange Multiplier (LM) unit root test to endogenously detect structural breaks to investigate the stationary properties of ELC per capita in 23 OECD countries selected in terms of high-income classifications. As a consequence of the test, it is claimed that the past behavior of the ELC per capita in almost all OECD countries will be used to forecast its future pattern because the unit root null hypothesis is rejected for 21 countries. Shahbaz et al. (2013) analyze whether the fluctuations in ELC per capita of 67 developed and developing countries are temporary or not, utilizing Lee and Strazicich's (2004) unit root test and Lagrange Multiplier (LM) test for 1971–2010. The evidence of the tests indicates that ELC per capita in 65 countries can return its trend path. Bolat et al. use the individual unit root test with structural breaks improved by Carrion et al. (2005), allowing for cross-sectional dependence and multiple structural breaks to determine the stationary properties of ELC per capita in 16 European countries for 1960–2009. The evidence of unit root test with intercept-no trend confirms that the null hypothesis of stationary can be accepted except for six countries comprising Belgium, France, Germany, Greece, Luxembourg, and Sweden. In contrast, the test result with intercept and trend affirms that the stationary of ELC seems to be not rejected except Luxembourg. All in All, it is concluded that the appearance of the shocks to ELC per capita is a temporary impact for 15 countries; in other words, ELC tends to return to its time trend. Khraief et al. (2016) apply a univariate and panel unit root test to analyze whether ELC in Sub-Saharan Africa countries contains a unit root for 1971–2013. In addition to the conventional panel unit root test involving FirstGPUT and SecGPUT, the LM panel unit root rest improved by Im et al. (2005) is also employed, and the model's result poses any events affecting ELC seem to be powerless. Dogan (2016) analyzes the appearance of the shocks on ELC utilized by sector in Turkey. The EL data is based on 12 regions of Turkey by four regions and total ELC by region involving 60 cases. The evidence indicates that 48 cases contain a unit root, which means that energy management can play a vital role in changing ELC.

Regarding the literature, great studies are trying to reach useful knowledge for the stationarity properties of the energy sources. In this study, we concentrate on the ELP from oil, gas, and coal (% of total) in six emerging countries considering data techniques like Basher et al. (2015), Tiwari and Albulescu (2016) by using newly panel unit root tests with sharp shifts and smooth breaks along with FirstGPUT and SecGPUT. The importance of EL for urban and industrialization needs for emerging countries and the more useful knowledge resulting from the econometric techniques may execute an essential role in expanding the existing literature.

3 Results

The result section covers the findings of the three types of the panel unit root tests: FirstGPUT and SecGPUT, and a panel root test that allows for considering both sharp and smooth breaks introduced Bahmani-Oskooee et al. (2014). All three types of panel unit root tests are employed to investigate whether the impacts of the shocks on ELP from NRER are temporary or permanent. Before conducting the panel stationarity tests, Cross-sectional dependence (CD) tests involving Breusch-Pagan LM, Pesaran Scaled LM, Bias-Corrected Scaled LM, and Pesaran CD (2004) are carried out. The result of the tests is reported in Table 1 Panel A. According to the finding, the existence of CD is found.

FirstGPUT involve Levin et al. (2002), Breitung (2001), Im et al. (2003), ADF-Fisher, and PP-fisher. The result of FirstGPUT tests is reported in Table 1 Panel B. According to Table 1 Panel B, it is concluded that the electricity production from oil, gas, and coal is not stationary; in other words, it contains a unit root. However, FirstGPUT tests are exposed to significant drawbacks without regarding the effects of CD. SecGPUT are an improved model of FirstGPUT because they consider the cross-sectional dependence. SecGPUT test improved by Pesaran (2004) is carried out in this study. The result poses that the effects of the shocks on ELP in emerging countries are permanent. FirstGPUT and SecGPUT imply that ELP from oil, gas, and coal in the emerging countries has unit roots; in other words, it is not stationary. The energy management policies aimed to reduce ELP from NRER will be effective.

Nevertheless, there is also some drawback related to SecGPUT. They do not consider structural breaks. Various studies focus on sharp structural breaks to investigate the properties of the fluctuations in the energy, but smooth breaks are received less attention by just limited studies. A newly proposed panel unit root test for sharp and smooth structural breaks introduced by Bahmani-Oskooee et al. (2014) is employed. The cross-sectional independent among the variables is required to conduct panel unit root tests considering structural breaks. Still, Table 1 for the result of the CD tests poses that there is a CD. Bahmani-Oskooee et al. (2014) offer that the method introduced by Maddala and Wu (1999)‘s producer is used to overcome this

Table 1 The finding of cross-sectional dependence and first-second generation panel unit root tests

Panel A: Cross-sectional dependence tests

Variables	Breusch-Pagan LM	Pesaran scaled LM	Bias-corrected scaled LM	Pesaran CD
Electricity production from oil, gas, and coal sources (% of total)	142.2798 (0.000)	22.14257 (0.000)	22.07439 (0 000)	8.826020 (0.000)

Panel B: First and second-generation panel unit root test

Variables	First-generation panel unit root test					Second-generation panel unit root test
	Common unit root process		Individual unit root process			Pesaran CADF
	Levin, Lin &Chu	Breitung	Lm, Pesaran and Shin	ADF-Fisher	PP-Fisher	
Electricity production from oil, gas, and coal sources (% of total)	1.89843 (0.9712)	2.66504 (0.9962)	1.56552 (0.9413)	8.75434 (0.7238)	15.1551 (0.2331)	0.836 (0.798)
D(Electricity production from oil, gas, and coal sources (% of total))	−7.78035 (0.000)	−5.16783 (0.000)	−8.94619 (0.000)	90.0122 (0.000)	257.175 (0.000)	−7.121 (0.000)

drawback. The bootstrap procedure of Maddala and Wu (1999) with 1000 replication is carried out to achieve the critical values. Therefore, the conclusion of the panel stationary test introduced by Bahmani-Oskooee et al. (2014) is reported in Table 2 Panel A. The KPSS statistics for the homogenous and heterogeneous tests are lower than the critical values at the 10% significance level, implying that the stationary null hypothesis for six emerging countries is accepted. In other words, ELP from oil, gas, and coal sources is stationary. Panel B in Table 2 presents the results of a univariate version of the stationary test. The critical values for the univariate version are measured using a Monte Carlo simulation with 1000 replication. Thus, the null hypothesis of stationarity is rejected at the 10% significance level for three countries: China, Indonesia, and India. In contrast, the null hypothesis of the stationarity test cannot be rejected for Brazil, Mexico, and Turkey. These results emphasize that the fluctuations in ELP or the effects of the shocks on ELP from NRER are temporary for Brazil, Mexico, and Turkey; whereas, the effects of the shocks on ELP from NRER are permanent for China, Indonesia, and India. According to the achieved results from all panel unit root tests, it can be argued that ELP from NRER contains unit roots as a result of FirstGPUT and SecGPUT. This result implies that energy management policies purposed to reduce NRER share seem to be effective. Nevertheless, the result of the panel stationarity tests with sharp shifts and smooth breaks indicates that ELP from NRER cannot respond to the energy policies aimed to diminish the share of the NRER in ELP because the result

Table 2 Panel unit root tests with multiple sharp breaks

Panel A: Panel unit root test							
Number of cross: 6.0000							
Time period: 1971–2015							
Number of replication: 1000							
Maximum number of breaks: 2.0000							
Maximum number of frequencies: 5.0000							
	Stat.	**90%**	**95%**	**97.5%**	**99%**		
Homogeneous Panel KPSS Test	−2.1445 (0.9840)	0.8993	1.2984	1.7724	2.2049		
Heterogeneous Panel KPSS Test	−2.0295 (0.9788)	−0.5500	−0.3030	−0.0497	0.3340		
Panel B: Univariate unit root test and multiple breaking dates							
	KPSS Test	**90%**	**95%**	**97.5%**	**99%**	**Multiple breaking dates**	
Brazil	0.0369	0.1128	0.1476	0.1889	0.2442	2011.0000	0.0000
China	0.1616	0.1499	0.1822	0.2204	0.2624	1990.0000	2011.0000
Indonesia	0.1021	0.1009	0.1333	0.1709	0.2065	1977.0000	1995.0000
India	0.1105	0.0813	0.0989	0.1309	0.1585	1984.0000	1994.0000
Mexico	0.1325	0.1432	0.1864	0.2278	0.2598	1977.0000	2000.0000
Turkey	0.0982	0.1609	0.2062	0.2658	0.3325	1974.0000	1998.0000
Panel C: The results for optimum frequency and f-statistic and its critical values							
	F-Stat	**90%**	**95%**	**97.5%**	**99%**		
Brazil (1)	24.7728	2.4486	3.2177	4.2783	5.5166		
China (4)	8.3743	2.4532	3.2528	4.0672	5.3204		
Indonesia (4)	6.1633	2.5167	3.3886	4.0680	5.3259		
India (1)	5.9975	2.4505	3.2250	4.2643	0.1585		
Mexico (4)	7.8619	2.4873	3.4247	4.1903	0.2598		
Turkey (4)	3.1885	2.5270	3.3166	4.2247	0.3325		

of the panel statistic for the homogenous and heterogeneous show that ELP from NRER is stationary. As for the univariate version of the stationary test, the temporary effects of the shocks on ELP holds for Brazil, Mexico, and Turkey; while, the permanent effect is valid for China, Indonesia, and India. The difference leads to unique energy management policies among the countries. The energy implementation used to diminish the share of the NRER in ELP is recommended for China, Indonesia, and India.

Besides, Panel C in Table 2 confirms that all F statistics belonging to countries are higher than the critical values, leading to the null hypothesis's rejection. Therefore, the trigonometric variables are significant, which implies that both the sharp and smooth breaks model can be used for all variables. The global energy crisis that occurred in the 1970s and the 1980s is experienced in Indonesia, India, Mexico, and Turkey due to the multiple breaking dates. However, China has experienced multiple Breaking Dates occurring in the 1990s and the 2000s, named the massive industrialization and the increase in the urban development in China. In these periods, energy demand for urban needs and industrialization in China seem to induce multiple breaks.

4 Conclusion

ELC has been one of the most crucial development indicators. Economic growth and development objectives place reliance on an adequate and stable supply of EL; all in all, EL is a requisite figure in every side of human life. Although sustainable access to EL becomes irreplaceable steps for economic development, types of energy sources providing EL have been another essential point. In the world, generating EL from NRER accounts for approximately 63.3% worldwide. NRER are the main culprit for environmental degradation experienced around the world. CO2 emissions, marine pollution, and habitat destruction are the detrimental effects of NRER, harming sustainable development goals. In addition, NRER are ultimately subordinate to run out of fatefulness. Regarding the damaging effects of NRER, EL generation from RER has become irreplaceable to achieve development objectives. Within this aim, the investigation for the stationary properties of ELP provides insight information for policymakers to design and implement energy policies. For example, if ELP generation from NRER contains unit roots, any shocks from energy policies will permanently impact ELP. However, if ELP does not involve unit root; in other explanation, the variable restores to its trend route in the aftereffect of shocks, the energy policies will have a transitory impact on ELP. In contrast, the presence of the stationary can be used in formulating forecasting. Within these objectives, the study investigates ELP from oil, gas, and coal (%of total) in six emerging countries to reach a piece of knowledge concerning stationarity characteristics of the considered variable.

Employing FirstGPUT and SecGPUT and a newly improved panel unit root test allowing for considering sharp shifts and smooth breaks introduced by

Bahmani-Oskooee et al. (2014) investigate the unit root characteristic of ELC from NRER (% of total) for six emerging countries over the period 1971–2015. The results of FirstGPUT and SecGPUT found that the ELP from NRER involves unit roots, which implies that shocks will have a power deviating the ELP from its long-run trend way. The policymakers can adopt defined target levels regarding the share of ELP from NRER. The energy management policies aimed to decrease the percentage of NRER in ELP will be effective. Nevertheless, the result of the panel unit root test introduced by Bahmani-Oskooee et al. (2014) indicates that homogenous and heterogeneous tests are not higher than the critical values, which presents the stationary hypothesis for six emerging countries. According to these implications, any shock-generating energy policies will temporarily impact ELP, which follows its long-run trend. Moreover, the univariate version of the stationary result for each country shows that the rejection of the null hypothesis holds for China, Indonesia, and India. In contrast, the acceptance of the null hypothesis is found for Brazil, Mexico, and Turkey. According to this finding, it is not recommended for six emerging countries to follow standard energy policies. Therefore, China, Indonesia, and India have a policy tool to diminish the share of NRER in ELP because the energy policies will have a long-lasting impression on electricity production.

References

Aydin, M., & Pata, U. K. (2020). Are shocks to disaggregated renewable energy consumption permanent or temporary for the USA? Wavelet based unit root test with smooth structural shifts. *Energy, 207*, 118245.

Bahmani-Oskooee, M., Chang, T., & Wu, T. (2014). Revisiting purchasing power parity in African countries: panel stationary test with sharp and smooth breaks. *Applied Financial Economics, 24*(22), 1429–1438.

Barros, C. P., Gil-Alana, L. A., & Payne, J. E. (2013). US disaggregated renewable energy consumption: persistence and long memory behavior. *Energy Economics, 40*, 425–432.

Basher, S. A., Masini, A., & Aflaki, S. (2015). Time series properties of the renewable energy diffusion process: implications for energy policy design and assessment. *Renewable and Sustainable Energy Reviews, 52*, 1680–1692.

Becker, R., Enders, W., & Lee, J. (2006). A stationarity test in the presence of an unknown number of smooth breaks. *Journal of Time Series Analysis, 27*(3), 381–409.

BP statistical review of World Energy. (2020). https://www.bp.com/content/dam/bp/business-sites/en/global/corporate/pdfs/energy-economics/statistical-review/bp-stats-review-2020-full-report.pdf

Breitung, J. (2001). *The local power of some unit root tests for panel data*. Emerald Group Publishing Limited.

Carrion, I. S., Del Barrio-Castro, T., & Lopez-Bazo, E. (2005). Breaking the panels: an application to GDP per capita. *The Econometrics Journal, 8*(2), 159–175.

Demir, E., & Gozgor, G. (2018). Are shocks to renewable energy consumption permanent or temporary? Evidence from 54 developing and developed countries. *Environmental Science and Pollution Research, 25*(4), 3785–3792.

Dogan, E. (2016). Are shocks to electricity consumption transitory or permanent? Sub-national evidence from Turkey. *Utilities Policy, 41*, 77–84.

Gozgor, G. (2016). Are shocks to renewable energy consumption permanent or transitory? An empirical investigation for Brazil, China, and India. *Renewable and Sustainable Energy Reviews, 66*, 913–919.

Hasanov, M., & Telatar, E. (2011). A re-examination of stationarity of energy consumption: evidence from new unit root tests. *Energy Policy, 39*(12), 7726–7738.

Hsu, Y. C., Lee, C. C., & Lee, C. C. (2008). Revisited: are shocks to energy consumption permanent or temporary? New evidence from a panel SURADF approach. *Energy Economics, 30*(5), 2314–2330.

Im, K. S., Pesaran, M. H., & Shin, Y. (2003). Testing for unit roots in heterogeneous panels. *Journal of Econometrics, 115*(1), 53–74.

Im, K. S., Lee, J., & Tieslau, M. (2005). Panel LM unit-root tests with level shifts. *Oxford Bulletin of Economics and Statistics, 67*(3), 393–419.

Işık, C., Ahmad, M., Ongan, S., Ozdemir, D., Irfan, M., & Alvarado, R. (2021). Convergence analysis of the ecological footprint: theory and empirical evidence from the USMCA countries. *Environmental Science and Pollution Research, 28*, 1–12.

Khraief, N., Omoju, O. E., & Shahbaz, M. (2016). Are fluctuations in electricity consumption per capita in Sub-Saharan Africa countries transitory or permanent? *Energy Strategy Reviews, 13*, 86–96.

Kula, F., Aslan, A., & Ozturk, I. (2012). Is per capita electricity consumption stationary? Time series evidence from OECD countries. *Renewable and Sustainable Energy Reviews, 16*(1), 501–503.

Lean, H. H., & Smyth, R. (2013). Are fluctuations in US production of renewable energy permanent or transitory? *Applied Energy, 101*, 483–488.

Lee, J., & Strazicich, M. C. (2004). *Minimum LM unit root test with one structural break (Working Paper)*. Department of Economics, Appalachian State University.

Levin, A., Lin, C. F., & Chu, C. S. J. (2002). Unit root tests in panel data: asymptotic and finite-sample properties. *Journal of Econometrics, 108*(1), 1–24.

Maddala, G. S., & Wu, S. (1999). A comparative study of unit root tests with panel data and a new simple test. *Oxford Bulletin of Economics and Statistics, 61*(S1), 631–652.

Mishra, V., & Smyth, R. (2014). Is monthly US natural gas consumption stationary? New evidence from a GARCH unit root test with structural breaks. *Energy Policy, 69*, 258–262.

Pesaran, M. H. (2004). *General diagnostic tests for cross section dependence in panels*. (IZA Discussion Paper No. 1240). Institute for the Study of Labor (IZA).

Shahbaz, M., Tiwari, A. K., Ozturk, I., & Farooq, A. (2013). Are fluctuations in electricity consumption per capita transitory? Evidence from developed and developing economies. *Renewable and Sustainable Energy Reviews, 28*, 551–554.

Shahbaz, M., Arouri, M., & Teulon, F. (2014). Short-and long-run relationships between natural gas consumption and economic growth: Evidence from Pakistan. *Economic Modelling, 41*, 219–226.

Shahbaz, M., Solarin, S., & Mallick, H. (2015). *Are Fluctuations in Gas Consumption Per Capita Transitory? Evidence from LM Unit Root Test with Two Structural Breaks*. University Library of Munich, Germany.

Smyth, R. (2013). Are fluctuations in energy variables permanent or transitory? A survey of the literature on the integration properties of energy consumption and production. *Applied Energy, 104*, 371–378.

Solarin, S. A., & Lean, H. H. (2016). Are fluctuations in oil consumption permanent or transitory? Evidence from linear and nonlinear unit root tests. *Energy Policy, 88*, 262–270.

Tang, X., Jin, Y., McLellan, B. C., Wang, J., & Li, S. (2018). China's coal consumption declining—Impermanent or permanent? *Resources, Conservation and Recycling, 129*, 307–313.

Tiwari, A. K., & Albulescu, C. T. (2016). Renewable-to-total electricity consumption ratio: estimating the permanent or transitory fluctuations based on flexible Fourier stationarity and unit root tests. *Renewable and Sustainable Energy Reviews, 57*, 1409–1427.

UNEP. (2015). *United Nations Environment Programme*. Annu. Rep. 2015. file:/UNEP_2015_Annual_Report-2016UNEP-AnnualReport-2015-EN.pdf.pdf.

Wang, Y., Li, L., Kubota, J., Zhu, X., & Lu, G. (2016). Are fluctuations in Japan's consumption of non-fossil energy permanent or transitory? *Applied Energy, 169*, 187–196.

Yilanci, V., & Tunali, Ç. B. (2014). Are fluctuations in energy consumption transitory or permanent? Evidence from a Fourier LM unit root test. *Renewable and Sustainable Energy Reviews, 36*, 20–25.

Financial Evaluation of Energy Investments in Russia

Elizaveta Ibragimova and Nora Baranyai

1 Introduction

Recovery from the mandatory energy Risk insurance Fund of the Agency's obligations arising in connection with the Agency's failure to fulfill its obligations to pay compensation for deposits is carried out only based on a judicial act. The Mandatory Energy Risk Insurance Fund is formed at the expense of the additional bid and the increased additional bid will be determined depending on several factors. Starting from January 1, 2016, in addition to this risk assessment component, the bank has introduced a financial stability component: the bank is required to pay a higher amount of insurance premium, the more its financial situation does not meet the criteria defined by Law.

Insurance premiums are payable by the bank from the date of entering the bank in the register of banks and until the day of revocation (cancellation) of the Bank of Russia license or until the day of exclusion of the bank from the register of banks. The introduction by the Bank of Russia of a moratorium on satisfaction of creditors' claims suspends the bank's obligation to pay insurance premiums for the duration of the specified moratorium. At the same time, the bank is obliged to pay insurance premiums for the settlement period during which the specified moratorium was introduced, including the day preceding the introduction of this moratorium. Billing period for payment of insurance premiums This is the calendar quarter of the year.

E. Ibragimova (✉)
Financial Research Institute of the Ministry of Finance of the Russian Federation, Moscow, Russia

N. Baranyai
Circular Economy University Center, Renewable Energy Research Group, University of Pannonia, Veszprém, Hungary

© The Author(s), under exclusive license to Springer Nature Switzerland AG 2022
H. Dinçer, S. Yüksel (eds.), *Sustainability in Energy Business and Finance*, Contributions to Finance and Accounting,
https://doi.org/10.1007/978-3-030-94051-5_5

To restore the mandatory energy risk insurance fund, the Agency's Board of Directors may raise the basic rate of insurance premiums. Insurance premiums are calculated by banks independently (Li et al., 2020; Yuan et al., 2021). Payment of insurance premiums is made within twenty-five days from the end of the billing period by transferring funds to the Agency's account with the Bank of Russia, which accounts for the funds of the mandatory energy risk insurance fund.

- Penalties for late and/or incomplete payment of insurance premiums. Recovery of monetary amounts of outstanding obligations of banks to pay insurance premiums, as well as penalties, is carried out by the Agency in court.
- Funds and other property received from the satisfaction of the Agency's claim rights acquired as a result of paying them compensation for deposits. To ensure the financial stability of the energy risk insurance system, the federal law on the federal budget for the corresponding year establishes the right of the Government of the Russian Federation to issue budget loans and borrow, the maximum amount of these borrowings, as well as the maximum amount of corresponding federal budget expenditures.
- Income from placement and (or) investment of temporarily available funds of the mandatory energy risk insurance fund (Fang et al., 2021; Qiu et al., 2020; Zhou et al., 2021). The directions, procedure, and conditions for placing and (or)-investing temporarily available funds of the mandatory energy risk insurance fund, as well as the maximum amount of temporarily available funds to be placed and (or)invested, are determined annually by the Agency's Board of Directors.

Temporarily available funds of the fund may not be used to purchase securities of issuers in respect of which pre-trial rehabilitation measures are being implemented, or bankruptcy proceedings have been initiated (supervision, temporary bankruptcy management, bankruptcy proceedings), or such procedures were applied during the previous two years.

- Initial property contribution.
- Other income that is not prohibited by the legislation of the Russian Federation. A loan from the Bank of Russia may be one of the sources of forming an energy risk insurance fund, as well as the Agency's implementation of measures to improve the financial health of banks.

2 Literature Review

If the Agency's Board of Directors is unable to reimburse deposits without adding additional funds to the mandatory energy risk insurance fund, the Agency's Board of Directors will make one of the following decisions within seven calendar days after the insured event:

- Apply to the Government of the Russian Federation with a request to allocate appropriate funds to the Agency in the form of a budget loan, if the deficit of the mandatory energy risk insurance fund calculated by the Agency's management board is not more than RUB 1 billion.

- Apply to the Agency's Board of Directors for compensation of deposits without adding additional funds to the mandatory energy risk insurance fund to the Government of the Russian Federation with a request to allocate additional funds to the Agency from the federal budget if the deficit of the mandatory energy risk insurance fund calculated by the Agency's management board exceeds 1 billion rubles (Bhuiyan et al., 2021; Dong et al., 2021; Dorofeev, 2020; Grilli et al., 2021; Liu et al., 2022; Mikhaylov, 2021b; Moiseev et al., 2021; Radosteva et al., 2018; Ranjbar et al., 2017; Rathnayaka et al., 2018; Saqib et al., 2021; Sunchalin et al., 2019; Udalov, 2021; Yüksel et al., 2021a; Yüksel et al., 2021b; Yüksel et al., 2021c).

The Agency also has the right to apply to the Bank of Russia for a loan to replenish the energy risk insurance fund for a period of up to five years. Control over the functioning of the energy risk insurance system is carried out by the Government of the Russian Federation and the Bank of Russia through the participation of their representatives in the Agency's management bodies (An et al., 2021; Conteh et al., 2021; Danish et al., 2020, 2021; Dayong et al., 2020; Ivanyuk, 2018; Ivanyuk et al., 2020; Ivanyuk & Berzin, 2020; Ivanyuk & Levchenko, 2020; Ivanyuk & Soloviev, 2019; Lisin, 2020; Mikhaylov et al., 2018; Nyangarika et al., 2018; Uandykova et al., 2020; Uyeh et al., 2021).

Let us consider in detail the key principles and definitions on which payments to affected citizens are based. These funds are placed by individuals in the selected banking institution, based on an agreement drawn up earlier between the project and the organization's representative (Kou et al., 2021; Silahtaroğlu et al., 2021). It is mandatory to include various percentages on the previously placed deposit amount as an insurance item (Jun et al., 2021; Liu et al., 2021a, 2021b; Melnichuk et al., 2020; Mikhaylov, 2018c; Mikhaylov et al., 2019; Mukhametov et al., 2021; Nie et al., 2020).

Items that do not relate to insured events include the following:

- Placement of foreign currency funds of individuals, in the framework of business activities, without the necessary legal entity formation. Energy risks fall into this category, provided that the necessary accounts are opened in full compliance with the activities carried out (Cheng et al., 2020; Haiyun et al., 2021; Liu et al., 2021a, 2021b; Zhe et al., 2021).
- In cases where funds are placed on a deposit in the name of the bearer, and in particular, if additionally, the data is verified by using a special certificate, or by a savings book also issued to bearer.
- If the placement of energy risks is carried out directly outside the territory of the Russian Federation, in various branches of Russian banks (Candila et al., 2021; Denisova et al., 2019; Huang et al., 2021a, 2021b; Meynkhard, 2019, 2020; Mikhaylov, 2018a, 2018b, 2022; Mikhaylov et al., 2019; Nyangarika et al., 2019a, 2019b).

The general list of insured events usually includes the following:

- Revocation or cancellation of the banking organization's existing license provided for performing subsequent operations and servicing potential customers.
- If the Central Bank of the Russian Federation imposes a moratorium on subsequent satisfaction of claims issued to creditors of a banking organization. In terms of time, the payment of compensation is carried out in accordance with the existing register of obligations of the liquidated bank to its projects. In particular, this is relevant within three days from the moment the project submits all the necessary documents to the Agency (Alwaelya et al., 2021; An & Mikhaylov, 2020, 2021; An et al., 2019a, 2019b, 2020a, 2020b, 2020c; Dooyum et al., 2020; Gura et al., 2020; Mikhaylov, 2020a, 2020b, 2020c, 2021a; Mikhaylov & Tarakanov, 2020; Mikhaylov et al., 2021a, 2021b; Moiseev et al., 2020; Morkovkin et al., 2020a, 2020b; Mutalimov et al., 2021; Varyash et al., 2020; Yumashev & Mikhaylov, 2020; Yumashev et al., 2020; Zhao et al., 2021).

However, this payment can be made no earlier than two weeks (14 calendar days) from the date of occurrence of the insured event provided for in the organization's regulations. By upon direct submission of the required set of documents to the Agency, it is issued a corresponding extract from the created register of obligations of the banking organization to its own projects, with a corresponding indication of the total amount of compensation for previously located deposits. So that users can be guided about the need to compensate for energy risks, it is planned to submit a corresponding application in the Bulletin of the Bank of Russia. The place, time, and order of admission are established in the relevant press directly at the bank's legal address. Applications and the corresponding data form defined by projects. Within one month from the moment of receipt of the required register of obligations from the banking organization, all necessary information is sent to projects on an individual basis at the direct request of the recipient.

3 Methods

The project, its successor or legal successor (their representatives) may apply to the Agency with the following request: a requirement on payment of compensation for deposits from the date of occurrence of the insured event to the date of completion of the insurance contract. Bankruptcy proceedings (forced liquidation), and if the Bank of Russia imposes a moratorium on satisfaction of creditors' claims—until the day when the moratorium expires. If the project (its successor, legal successor) misses the deadline for filing a claim for compensation for deposits, the deadline for submitting a claim for compensation for deposits is set by the project (its successor, legal successor) to the application form the project (its successor, legal successor) may be restored by a decision of the Agency's Management Board in the presence of one of the following circumstances:

- If the application of the project (its successor, legal successor) with a claim for payment of compensation for deposits was prevented by an extraordinary and unavoidable circumstance under these conditions.
- If the project (heir) was (is) undergoing military service on conscription or was (is) a member of the Armed Forces of the Russian Federation (other troops, military formations, bodies) transferred to martial law, for the period of such service (martial law).
- If the reason for missing the specified deadline is related to the serious illness of the project (its heir), the helpless state of the project (its heir), the deadline for the project heir to accept the inheritance, and other reasons related to the personality of the project (its heir).

The decision of the Agency's Management Board to refuse to restore the missed deadline for filing a claim for payment of compensation for deposits may be appealed the project (its successor, legal successor) to the court.

When applying to the Agency with a claim for payment of compensation for deposits, the project, the heir, the legal successor (their representatives) submit: an application for compensation for deposits in the amount of in the form of if an individual applies, documents confirming his/her identity; if an heir applies, documents confirming his/her right to inheritance or the right to use the testator's funds; if a legal successor applies, documents confirming the transfer of the right to claim a deposit to him/her; if a project representative, heir's representative, or legal successor applies, a notarized power of attorney (except for a person authorized to act on behalf of the project, heir, or legal successor without a power of attorney).

Payment of the actual refund is made directly on the application given by the potential project, which is carried out both by transferring cash and by transferring the requested amount to the account of another bank specified in the submitted documents by the project. Acceptance of this application, as well as other documents in the set required for obtaining material resources, is carried out by the Agency through various agent banks that act directly on behalf of the project and strictly at its expense. The refund is made for all deposits with the banking organization, if they occur in the specified case, in the amount of at least 100% of the current deposit amount. However, the maximum amount of compensation is also set, no more than 1,400,000 rubles. If the project has several separate energy risks on its account, this refund is paid pro rata for each of the specific bank energy risks. If the funds were in a foreign currency, therefore, when making compensation payments, they are converted into rubles, at the current exchange rate.

4 Results

Therefore, in whatever currency the deposit is placed, payments are made only in the national currency, namely in rubles of the Russian Federation. Special attention should be paid to reimbursement of funds in a situation where the bank acted as a lender in relation to a specific project. The amount of compensation in this case is calculated based on the difference between the current amount of debt obligations to the project and the established amount of the counterclaim. To apply, the project will need to contact the appropriate Agency, or any selected agent bank, if it is established that it is involved in subsequent payments of the established compensation for previously made deposits. The right to do so is valid from the immediate day of the occurrence of the case, until the nominal moment when the bank's bankruptcy is considered to have taken place. The same applies to cases when the Central Bank of Russia establishes a moratorium.

A separate insurance indemnity, the maximum amount of which is up to ten million rubles, is paid:

- On an escrow account opened for settlements under a real estate purchase and sale transaction.
- On the escrow account opened for settlements under the contract of participation in shared capital construction.
- When funds are credited to the project's accounts from the sale of an apartment, residential building and land plot under it, garden house, and other buildings.
- Inherited funds, insurance, and social payments, as well as benefits and compensations, funds transferred by a court decision, grants in the form of subsidies.

Account holders, bank clients who have a power of attorney to receive insurance payments, as well as individual entrepreneurs, small and medium-sized businesses are entitled to insurance payments. It is important that they are listed in the unified register and heirs at the time of the insured event. If the client has several deposits in different banks, the refund is paid for each deposit in proportion to their size. If the deposit was made in dollars, euros, yen, etc., the currency will be converted into rubles at the exchange rate of the Central Bank on the day of the insured event and compensation will be paid.

There are cases when a client opens several different types of energy risks in the same bank. What should I do with paying a refund in this situation? The Central Bank will count these deposits as one deposit, and recalculate the compensation based on their total amount. If the deposits were placed in the bank's branches, the same measures are applied to them. Therefore, it is worth placing deposits in different banks, so as not to face such a deplorable situation.

Further, we note the problems of the Russian energy risk insurance system. In modern conditions, the actions of the Central Bank of the Russian Federation, which are aimed at "improving" the banking sector, lead to an increase in insurance claims. The number of banks whose licenses are being revoked by the Bank of Russia is constantly growing. The process of forming a fund takes time, while the Bank of Russia implements an active policy of revoking licenses. Currently, there is only one way to solve this problem—lending by the Bank of Russia to the energy Agency under special conditions. Also, a problem is that not all commercial banks are included in the energy risk insurance system, thus projects have risks of non-return of funds in the event of an insured event.

Currently, there are many problems that are associated with fraud in the field of energy risk insurance: fragmentation of energy risks, artificial formation of energy risks to illegally obtain insurance compensation (fictitious energy risks). In Russia, there is no deposit insurance for legal entities, which leads to significant losses of corporate funds. This problem can be solved in several ways: the introduction of mandatory deposit insurance for the corporate sector within certain limits (like energy risk insurance for individuals) and the development of voluntary deposit insurance for the corporate sector. Within the framework of the general financial illiteracy of the population in the Russian Federation, there is also a problem of illiteracy of citizens in the field of insurance, including energy risk insurance.

The high interest rates that attract investors often hide risks that are associated with financial instability, and subsequently the upcoming bankruptcy of a commercial bank. Clients have little knowledge of the real financial situation of credit institutions. So, the financial basis of energy risk insurance is the foundation of the mandatory energy risk insurance (Fund), at the expense of which compensation payments are made for deposits and expenses related to by an organization payout.

5 Conclusions and Discussion

In today's market economy, one of the keyways to save money is a bank deposit. Through energy risks, temporarily available funds are mobilized in the banking system and their further transformation into productive investments, which makes it possible to provide consumer loans to the population and satisfies the banking system's need for fixed and working capital. To maintain socio-economic interests and a high level of public confidence in this method of saving in most major countries, there is an energy risk insurance system that ensures the stability of the banking system and public confidence in financial and credit institutions.

In a market economy, large financial and credit organizations are required to ensure the safety of energy risks and timely fulfillment of their obligations to projects. The return of citizens' energy risks through compulsory insurance is guaranteed in the event of revocation (cancellation) of the Central Bank of the Russian Federation of the license of the bank where the deposit was placed or after the introduction of a moratorium on meeting the claims of other creditors of the bank. Since 2003, energy risk insurance for individuals in banks of the Russian Federation has been provided by the Energy Insurance Agency.

Commercial banks regularly pay insurance premiums in the amount of 0.1% of all deposits to energy agencies. Thus, customers do not personally make an additional payment for deposit insurance, and this obligation is performed by the bank at the basic, additional, or increased additional rate based on the current legislation. According to the legislation of the Energy Risk Insurance Agency, funds placed on the accounts of individuals of a commercial bank are subject to insurance, and an insured event on deposits with the bank occurs after the relevant decision is made by the Central Bank of the Russian Federation.

References

Alwaelya, S. A., Yousif, N. B. A., & Mikhaylov, A. (2021). Emotional development in preschoolers and socialization. *Early Child Development and Care, 190*, 2484–2493.
An, J., & Mikhaylov, A. (2020). Russian energy projects in South Africa. *Journal of Energy in Southern Africa, 31*(3), 336–342.
An, J., & Mikhaylov, A. (2021). Analysis of energy projects financial efficiency and renewable energy generation in Russia. *Finance: Theory and Practice, 25*(5), 1–11.
An, J., Mikhaylov, A., & Sokolinskaya, N. (2019a). Machine learning in economic planning: Ensembles of algorithms. *Journal of Physics: Conference Series, 1353*, 012126.
An, J., Mikhaylov, A., & Moiseev, N. (2019b). Oil price predictors: Machine learning approach. *International Journal of Energy Economics and Policy, 9*(5), 1–6.
An, J., Mikhaylov, A., & Richter, U. H. (2020a). Trade war effects: Evidence from sectors of energy and resources in Africa. *Heliyon, 6*, e05693. https://doi.org/10.1016/j.heliyon.2020.e05693
An, J., Mikhaylov, A., & Jung, S.-U. (2020b). The strategy of South Korea in the global oil market. *Energies, 13*(10), 2491.
An, J., Mikhaylov, A., & Kim, K. (2020c). Machine learning approach in heterogeneous group of algorithms for transport safety-critical system. *Applied Sciences, 10*(8), 2670.
An, J., Mikhaylov, A., & Jung, S.-U. (2021). A linear programming approach for robust network revenue management in the airline industry. *Journal of Air Transport Management, 91*, 101979.
Bhuiyan, M. A., An, J., Mikhaylov, A., Moiseev, N., & Danish, M. S. S. (2021). Renewable energy deployment and COVID-19 measures for sustainable development. *Sustainability, 13*, 4418. https://doi.org/10.3390/su13084418
Candila, V., Maximov, D., Mikhaylov, A., Moiseev, N., Senjyu, T., & Tryndina, N. (2021). On the relationship between oil and exchange rates of oil-exporting and oil-importing countries: From the great recession period to the Covid-19 era. *Energies, 14*, 8065.
Cheng, F., Lin, M., Yüksel, S., Dinçer, H., & Kalkavan, H. (2020). A hybrid hesitant 2-tuple IVSF decision making approach to analyze PERT-based critical paths of new service development process for renewable energy investment projects. *IEEE Access, 9*, 3947–3969.
Conteh, F., Takahashi, H., Hemeida, A. M., Krishnan, N., Mikhaylov, A., & Senjyu, T. (2021). Hybrid grid-connected renewable power generation for sustainable electricity supply in Sierra Leone: Case study Lungi Town. *Sustainability, 13*(18), 11435.
Danish, M. S. S., Bhattacharya, A., Stepanova, D., Mikhaylov, A., Grilli, M. L., Khosravy, M., & Senjyu, T. (2020). A systematic review of metal oxide applications for energy and environmental sustainability. *Metals, 10*(12), 1604.
Danish, M. S. S., Senjyu, T., Sabory, N. R., Khosravy, M., Grilli, M. L., Mikhaylov, A., & Majidi, H. (2021). A forefront framework for sustainable aquaponics modeling and design. *Sustainability, 13*, 9313. https://doi.org/10.3390/su13169313
Dayong, N., Mikhaylov, A., Bratanovsky, S., Shaikh, Z. A., & Stepanova, D. (2020). Mathematical modeling of the technological processes of catering products production. *Journal of Food Process Engineering, 43*(2), 1–9.
Denisova, V., Mikhaylov, A., & Lopatin, E. (2019). Blockchain infrastructure and growth of global power consumption. *International Journal of Energy Economics and Policy, 9*(4), 22–29.
Dong, B., Ikonnikova, I., Rogulin, R., Sakulyeva, T., & Mikhaylov, A. (2021). Environmental-economic approach to optimization of transport communication in megacities. *Journal of Environmental Science and Health, Part A, 56*(6), 660–666. https://doi.org/10.1080/10934529.2021.1913928
Dooyum, U. D., Mikhaylov, A., & Varyash, I. (2020). Energy security concept in Russia and South Korea. *International Journal of Energy Economics and Policy, 10*(4), 102–107.
Dorofeev, M. L. (2020). Impact of monetary policy on the level of economic inequality in the United States. *MGIMO Review of International Relations, 13*(5), 97–114. https://doi.org/10.24833/2071-8160-2020-5-74-97-114

Fang, S., Zhou, P., Dinçer, H., & Yüksel, S. (2021). Assessment of safety management system on energy investment risk using house of quality based on hybrid stochastic interval-valued intuitionistic fuzzy decision-making approach. *Safety Science, 141*, 105333.

Grilli, M. L., Valerini, D., Rizzo, A., Yilmaz, M., Song, C., Hu, G., Chierchia, R., Mikhaylov, A., & Rinaldi, A. (2021). A comparative study of the mechanical and tribological properties of thin Al2O3 coatings fabricated by atomic layer deposition and radiofrequency sputtering. *Physica Status Solidi (A), 218*(24), 1–25.

Gura, D., Mikhaylov, A., Glushkov, S., Zaikov, M., & Shaikh, Z. A. (2020). Model for estimating power dissipation along the interconnect length in single on-chip topology. *Evolutionary Intelligence*, s12065. https://doi.org/10.1007/s12065-020-00407-7

Haiyun, C., Zhixiong, H., Yüksel, S., & Dinçer, H. (2021). Analysis of the innovation strategies for green supply chain management in the energy industry using the QFD-based hybrid interval valued intuitionistic fuzzy decision approach. *Renewable and Sustainable Energy Reviews, 143*, 110844.

Huang, Y., Masrur, H., Shigenobu, R., Hemeida, A. M., Mikhaylov, A., & Senjyu, T. (2021a). A comparative design of a campus microgrid considering a multi-scenario and multi-objective approach. *Energies, 14*, 2853. https://doi.org/10.3390/en14112853

Huang, Y., Yona, A., Takahashi, H., Hemeida, A. M., Mandal, P., Mikhaylov, A., Senjyu, T., & Lotfy, M. E. (2021b). Energy management system optimization of drug store operate electric vehicles Charging Station. *Sustainability, 14*, 1217836.

Ivanyuk, V. (2018). Econometric forecasting models based on forecast combination methods. In *Proceedings of 2018 11th International Conference Management of Large-Scale System Development*, MLSD 2018. https://doi.org/10.1109/MLSD.2018.8551825.

Ivanyuk, V., & Berzin, D. (2020). An algorithm for constructing an efficient investment portfolio. In R. Silhavy, P. Silhavy, & Z. Prokopova (Eds.), *Software engineering perspectives in intelligent systems. CoMeSySo 2020. Advances in intelligent systems and computing* (Vol. 1294). Springer. https://doi.org/10.1007/978-3-030-63322-6_39

Ivanyuk, V., & Levchenko, K. (2020). Intelligent methods for predicting financial time series. In R. Silhavy, P. Silhavy, & Z. Prokopova (Eds.), *Software engineering perspectives in intelligent systems. CoMeSySo 2020. Advances in intelligent systems and computing* (Vol. 1294). Springer. https://doi.org/10.1007/978-3-030-63322-6_41

Ivanyuk, V., Soloviev, V. (2019). Efficiency of neural networks in forecasting problems. In *Proceedings of 2019 12th International Conference Management of Large-Scale System Development*; MLSD 2019. https://doi.org/10.1109/MLSD.2019.8911046

Ivanyuk, V., Sunchalin, A., & Sunchalina, A. (2020). Development of an intelligent ensemble forecasting system. In R. Silhavy, P. Silhavy, & Z. Prokopova (Eds.), *Software engineering perspectives in intelligent systems. CoMeSySo 2020. Advances in intelligent systems and computing* (Vol. 1294). Springer. https://doi.org/10.1007/978-3-030-63322-6_40

Jun, Q., Dinçer, H., & Yüksel, S. (2021). Stochastic hybrid decision-making based on interval type 2 fuzzy sets for measuring the innovation capacities of financial institutions. *International Journal of Finance & Economics, 26*(1), 573–593.

Kou, G., Akdeniz, Ö. O., Dinçer, H., & Yüksel, S. (2021). Fintech investments in European banks: A hybrid IT2 fuzzy multidimensional decision-making approach. *Financial Innovation, 7*(1), 1–28.

Li, X., Zhu, S., Yüksel, S., Dinçer, H., & Ubay, G. G. (2020). Kano-based mapping of innovation strategies for renewable energy alternatives using hybrid interval type-2 fuzzy decision-making approach. *Energy, 211*, 118679.

Lisin, A. (2020). Valuation of the activities of foreign banks in the Russian banking sector. *Orbis, 15*(45), 53–63. http://www.revistaorbis.org.ve/pdf/45/art5.pdf

Liu, Y., Gong, X., Yüksel, S., Dinçer, H., & Aydın, R. (2021a). A multidimensional outlook to energy investments for the countries with continental shelf in East Mediterranean region with hybrid decision making model based on IVIF logic. *Energy Reports, 7*, 158–173.

Liu, Z.-J., Panfilova, E., Mikhaylov, A., & Kurilova, A. (2021b). COVID-19 crisis impact on the stability between parties in crowdfunding and crowdsourcing. *Wireless Personal Communications, 122*(1), 915–930. https://doi.org/10.1007/s11277-021-08932-z

Liu, Z.-J., Panfilova, E., Mikhaylov, A., & Kurilova, A. (2022). Assessing stability in relationship between parties in crowdfunding and crowdsourcing projects during the COVID-19 crisis. *Journal of Global Information Management, 30*(2), 1.

Melnichuk, A. V., Melnichuk, Y. A., Bocharov, A. V., Stepanova, D. I., & Kurbanov, S. A. (2020). Constitutional foundations of the functioning of local self-government. *Revista Inclusiones., 7*, 337–346.

Meynkhard, A. (2019). Fair market value of bitcoin: Halving effect. *Investment Management and Financial Innovations, 16*(4), 72–85. https://doi.org/10.21511/imfi.16(4).2019.07

Meynkhard, A. (2020). Effect of bitcoin volatility on altcoins pricing. *Advances in Intelligent Systems and Computing, 1294*, 652–664. https://doi.org/10.1007/978-3-030-63322-6_55

Mikhaylov, A. (2018a). Volatility spillover effect between stock and exchange rate in oil exporting countries. *International Journal of Energy Economics and Policy, 8*(3), 321–326.

Mikhaylov, A. (2018b). Pricing in oil market and using probit model for analysis of stock market effects. *International Journal of Energy Economics and Policy, 8*(2), 69–73.

Mikhaylov, A. (2018c). The yield of the carry trade strategy. *Finance: Theory and Practice, 22*(3), 52–63. https://doi.org/10.26794/2587-5671-2018-22-3-52-63

Mikhaylov, A. (2020a). Cryptocurrency market analysis from the open innovation perspective. *Journal of Open Innovation: Technology, Market, and Complexity, 6*(4), 197. https://doi.org/10.3390/joitmc6040197

Mikhaylov, A. (2020b). Geothermal energy development in Iceland. *International Journal of Energy Economics and Policy, 10*(4), 31–35.

Mikhaylov, A. (2020c). Cryptocurrency market development: Hurst method. *Finance: Theory and Practice, 24*(3), 81–91.

Mikhaylov, A. (2021a). Development of Friedrich von Hayek's theory of private money and economic implications for digital currencies. *Terra Economicus, 19*(1), 53–62.

Mikhaylov, A. (2021b). Lichens as indicators of atmospheric pollution in urban ecosystems. *Israel Journal of Ecology & Evolution, 67*(1–2), 60–68. https://doi.org/10.1163/22244662-bja10016

Mikhaylov, A. (2022). Efficiency of renewable energy generation in Russia. *Annals of the Brazilian Academy of Sciences, 94*(3), 1.

Mikhaylov, A., & Tarakanov, S. (2020). Development of Levenberg-Marquardt theoretical approach for electric network. *Journal of Physics: Conference Series, 1515*, 052006.

Mikhaylov, A., Sokolinskaya, N., & Nyangarika, A. (2018). Optimal carry trade strategy based on currencies of energy and developed economies. *Journal of Reviews on Global Economics, 7*, 582–592.

Mikhaylov, A., Burakov, D., & Didenko, V. (2019). Relationship between oil Price and macroeconomic indicators in Russia. *Finance: Theory and Practice, 23*(2), 105–116. https://doi.org/10.26794/2587-5671-2019-23-2-105-116

Mikhaylov, A., Yumashev, A., & Kolpak, E. (2021a). Quality of life, anxiety and depressive disorders in patients with extrasystolic arrhythmia. *Archives of Medical Science, 17*(1). https://doi.org/10.5114/aoms.2020.101359

Mikhaylov, A., Danish, M. S. S., & Senjyu, T. (2021b). New stage in evolution of cryptocurrency market: Analysis by Hurst method. In H. Dincer & S. Yuksel (Eds.), *Strategic outlook in business and finance innovation: Multidimensional policies for emerging economies* (pp. 35–45. ISBN: 978-1-80043-445-5). Emerald Publishing Limited. https://doi.org/10.1108/978-1-80043-444-820211004

Moiseev, N., Mikhaylov, A., Varyash, I., & Saqib, A. (2020). Investigating the relation of GDP per capita and corruption index. *Entrepreneurship and Sustainability Issues, 8*(1), 780–794.

Moiseev, N., Sorokin, A., Zvezdina, N., Mikhaylov, A., Khomyakova, L., & Danish, M. S. S. (2021). Credit risk theoretical model on the base of DCC-GARCH in time-varying parameters framework. *Mathematics, 9*, 2423. https://doi.org/10.3390/math9192423

Morkovkin, D. E., Gibadullin, A. A., Kolosova, E. V., Semkina, N. S., & Fasehzoda, I. S. (2020a). Modern transformation of the production base in the conditions of industry 4.0: Problems and prospects. *Journal of Physics: Conference Series, 1515*, 032014.

Morkovkin, D., Lopatkin, D., Sadriddinov, M., Shushunova, T., Gibadullin, A., & Golikova, O. (2020b). Assessment of innovation activity in the countries of the world. *E3S Web of Conferences, 157*, 04015. https://doi.org/10.1051/e3sconf/202015704015

Mukhametov, A., Bekhorashvili, A., Avdeenko, A., & Mikhaylov, A. (2021). The impact of growing legume plants under conditions of biologization and soil cultivation on chernozem fertility and productivity of rotation crops. *Legume Research, 44*, 12.

Mutalimov, V., Kovaleva, I., Mikhaylov, A., & Stepanova, D. (2021). Assessing regional growth of small business in Russia. *Entrepreneurial Business and Economics Review, 9*(3), 119–133. https://doi.org/10.15678/EBER.2021.090308

Nie, D., Panfilova, E., Samusenkov, V., & Mikhaylov, A. (2020). E-learning financing models in Russia for sustainable development. *Sustainability, 12*(11), 4412.

Nyangarika, A., Mikhaylov, A., & Tang, B.-J. (2018). Correlation of oil prices and gross domestic product in oil producing countries. *International Journal of Energy Economics and Policy, 8*(5), 42–48.

Nyangarika, A., Mikhaylov, A., & Richter, U. (2019a). Influence oil price towards economic indicators in Russia. *International Journal of Energy Economics and Policy, 9*(1), 123–129.

Nyangarika, A., Mikhaylov, A., & Richter, U. (2019b). Oil price factors: Forecasting on the base of modified auto-regressive integrated moving average model. *International Journal of Energy Economics and Policy, 9*(1), 149–159.

Qiu, D., Dinçer, H., Yüksel, S., & Ubay, G. G. (2020). Multi-faceted analysis of systematic risk-based wind energy investment decisions in E7 economies using modified hybrid modeling with IT2 fuzzy sets. *Energies, 13*(6), 1423.

Radosteva, M., Soloviev, V., Ivanyuk, V., & Tsvirkun, A. (2018). Use of neural network models in market risk management. *Advances in Systems Science and Applications, 18*(2), 53–58. https://doi.org/10.25728/assa.2018.18.2.582

Ranjbar, O., Chang, T., Nel, E., & Gupta, R. (2017). Energy consumption and economic growth nexus in South Africa: Asymmetric frequency domain approach. *Energy Sources, Part B: Economics, Planning, and Policy, 12*(1), 24–31.

Rathnayaka, R. K. T., Seneviratna, D. M. K., & Long, W. (2018). The dynamic relationship between energy consumption and economic growth in China. *Energy Sources, Part B: Economics, Planning, and Policy, 13*(5), 264–268.

Saqib, A., Chan, T.-H., Mikhaylov, A., & Lean, H. H. (2021). Are the responses of sectoral energy imports asymmetric to exchange rate volatilities in Pakistan? Evidence from recent foreign exchange regime. *Frontiers in Energy Research, 9*, 614463. https://doi.org/10.3389/fenrg.2021.614463

Silahtaroğlu, G., Dinçer, H., & Yüksel, S. (2021). Defining the significant factors of currency exchange rate risk by considering text mining and fuzzy AHP. In *Data science and multiple criteria decision making approaches in finance* (pp. 145–168). Springer.

Sunchalin, A. M., Kochkarov, R. A., Levchenko, K. G., Kochkarov, A. A., & Ivanyuk, V. A. (2019). Methods of risk management in portfolio theory. *Espacios, 40*(16), 8.

Uandykova, M., Lisin, A., Stepanova, D., Baitenova, L., Mutaliyeva, L., Yuksel, S., & Dincer, H. (2020). The social and legislative principles of counteracting ransomware crime. *Entrepreneurship and Sustainability Issues, 8*(2), 777–798.

Udalov, I. (2021). The transition to renewable energy sources as a threat to resource economies. *International Journal of Energy Economics and Policy, 11*(3), 1–6.

Uyeh, D. D., Asem-Hiablie, S., Park, T., Kim, K.-M., Mikhaylov, A., Woo, S., & Ha, Y. (2021). Could Japonica Rice be an alternative variety for increased global food security and climate change mitigation? *Food, 10*, 1869. https://doi.org/10.3390/foods10081869

Varyash, I., Mikhaylov, A., Moiseev, N., & Aleshin, K. (2020). Triple bottom line and corporate social responsibility performance indicators for Russian companies. *Entrepreneurship and Sustainability Issues, 8*(1), 313–331.

Yuan, G., Xie, F., Dinçer, H., & Yüksel, S. (2021). The theory of inventive problem solving (TRIZ)-based strategic mapping of green nuclear energy investments with spherical fuzzy group decision-making approach. *International Journal of Energy Research, 45*(4), 6435.

Yüksel, S., Khomyakova, L., & Mikhaylov, A. (2021a). Energy center selection in G7 industry with fuzzy MOORA. In S. Yüksel & H. Dinçer (Eds.), *Handbook of research on strategic management for current energy investments* (pp. 87–106). IGI Global. ISBN13: 9781799883357. https://doi.org/10.4018/978-1-7998-8335-7.ch006

Yüksel, S., Mikhaylov, A., Ubay, G. G., & Uyeh, D. D. (2021b). The role of hydrogen in the Black Sea for the future energy supply security of Turkey. In S. Yüksel & H. Dinçer (Eds.), *Handbook of research on strategic management for current energy investments* (pp. 1–15). IGI Global). https://doi.org/10.4018/978-1-7998-8335-7.ch001

Yüksel, S., Mikhaylov, A., & Ubay, G. G. (2021c). Factors causing delay in the installation of nuclear power plants. In S. Yüksel & H. Dinçer (Eds.), *Strategic approaches to energy management. contributions to management science*. Springer. https://doi.org/10.1007/978-3-030-76783-9_7

Yumashev, A., & Mikhaylov, A. (2020). Development of polymer film coatings with high adhesion to steel alloys and high wear resistance. *Polymer Composites, 41*(7), 2875–2880.

Yumashev, A., Ślusarczyk, B., Kondrashev, S., & Mikhaylov, A. (2020). Global indicators of sustainable development: Evaluation of the influence of the human development index on consumption and quality of energy. *Energies, 13*, 2768.

Zhao, Q., Cherkasov, S., Avdeenko, A., Kondratenko, L., & Mikhaylov, A. (2021). Integral estimate of the added value of agricultural crops in the synergetic agriculture on the example of Vicia faba (Vicia faba L.). *Polish Journal of Environmental Studies, 30*(5), 1–9. https://doi.org/10.15244/pjoes/131949

Zhe, L., Yüksel, S., Dinçer, H., Mukhtarov, S., & Azizov, M. (2021). The positive influences of renewable energy consumption on financial development and economic growth. *SAGE Open, 11*(3), 21582440211040133.

Zhou, P., Luo, J., Cheng, F., Yüksel, S., & Dinçer, H. (2021). Analysis of risk priorities for renewable energy investment projects using a hybrid IT2 hesitant fuzzy decision-making approach with alpha cuts. *Energy, 224*, 120184.

Strategic Talent Perception in the Energy Sector

Gizem Topsakal Acet and Pelin Vardarlıer

1 Introduction

Globalization and today's competitive environment have brought the importance of the internal resources of organizations to the forefront much more. As the resource-based view demonstrates, businesses need to focus on internal resources and manage them effectively and efficiently to provide competitive advantage. One of the most important internal resources of organizations is human. It is also important to place the right people at the right time and in the right place to achieve strategic goals and survive in a competitive environment in the long term (Al Ariss et al., 2014). On the other hand, finding and attracting talented employees has become difficult for organizations. However, organizations also face problems in creating a suitable business environment for the new talented generation (Farndale et al., 2010). Recruitment and talent acquisition can be compared short term and long term. Both approaches can be used depending on the circumstances. However, by its recruitment nature, tactics tend to be strategic in acquiring talent (Sparrow & Makram, 2015). For this reason, the concept of strategic talent stands out.

Strategic talent management has become a key business activity and a critical decision area for managers due to the lack of talent available in businesses. Strategic talent management requires strategic focus on its own. Therefore, it is difficult for any business management to achieve success in the long run, regardless of what kind of capabilities the business needs. Therefore, strategic talent management needs to be maintained by businesses (Sparrow & Makram, 2015). In addition, it is important

G. T. Acet
Bahçeşehir University, Istanbul, Turkey

P. Vardarlıer (✉)
School of Business, İstanbul Medipol University, Istanbul, Turkey
e-mail: pvardarlier@medipol.edu.tr

© The Author(s), under exclusive license to Springer Nature Switzerland AG 2022
H. Dinçer, S. Yüksel (eds.), *Sustainability in Energy Business and Finance*, Contributions to Finance and Accounting,
https://doi.org/10.1007/978-3-030-94051-5_6

for businesses to acquire talented employees, to keep these employees in their hands and to replace those who quit their jobs to make the competitive advantage of the business sustainable. In this context, businesses should attach the necessary importance to strategic talent management and integrate it with human resources processes. Accordingly, the aim of the research is to examine the perspectives of the enterprises on strategic talent management and to determine the level of these strategies in the enterprises.

Within the scope of this study, it focused on the field of energy sector. When we look at the researches analyzing the current situation of the energy sector, it is seen that one of the most important problems in this field is the inadequacy of institutional resources and human capacity (Bogomolova, 2018; Hussaini & Majid, 2014; Markovska et al., 2009). There are two views on the role of human resources in the sustainable development of the energy sector. One is the impact of the energy sector on the social dimension, and the other is the impact of human resources on know-how technologies and innovation in the field of energy (Angheluță et al., 2014). Accordingly, it is important to manage the right human resources for the energy sector. Indeed, the lack of long-term investment in the new energy capacity has led to a lack of training and transfer of technical knowledge in the development of qualified workforce in the energy sector (Markovska et al., 2009). Sustainability is one of the most prominent issues in the energy sector (Qi et al., 2020). At this point, it should be sustainable in its capabilities. It will be useful to ensure that people acquire the right skills through vocational training programs in the development of human capacity (Muda et al., 2017; Smits et al., 2014). Therefore, the importance of developing strategic capabilities in enterprises is critical. After identifying the perspectives of enterprises on strategic talent perception, suggestions for the subject were included.

2 Literature Review

Strategic talent management research has been reviewed to provide important information about under-discovered themes, topics, theories, and methods in the field of strategic talent management. In this context, Shet (2020) mentioned the perspectives of organizational stakeholders on strategic talent management. With the study, it was concluded that strategic talent management practices in organizations are beneficial in achieving the goals of organizations. In the study conducted by Borisova et al. (2017), the main approaches to the adoption of strategic talent management approach in managers were examined theoretically. According to the results of the research, it was seen that the implementation of strategic talent management in companies would enable companies to increase the efficiency of their motivation programs.

Karadal (2019), on the other hand, discussed the concept of talent management at the theoretical level, which brings a new vision to strategic human resources management. As a result of the research, it was determined that the organizations that apply talent management managed to be different from their competitors with the confidence in the products they produce and the services they provide. According to the results of the study conducted by Anlesinya et al. (2019), strategic talent management can be used to achieve positive results at the organizational and macro level. However, the realization of these positive outcomes is that talent management strategies can be threatened with a variety of challenges that need to be addressed in fulfilling critical conditions for their success. Accordingly, it is possible to say that the success of an effective talent management strategy should be under the collective responsibility of multiple stakeholders, not on the shoulders of a single individual.

In another study, Schreuder and Noorman (2019) argued that traditional strategic talent management practices cannot lead to organizational excellence. As a result of the research, it was determined that strategic talent management should harmonize and mutually strengthen business development and personal development to increase strategic success. Sripirom et al. (2016) found that employee talent development concentration had the strongest positive significant effect on strategic talent management. In another study conducted by Benoy and Gracias (2015), how businesses can improve the skills of existing project managers and use their experience and expertise to create a future talent pool was examined. According to the results of the research, it was stated that the lack of strategic skills in managers may lead to a lack of motivation among employees, delay or failure of projects, and may cause high monetary losses in the organization.

In the study conducted by Chen et al. (2021), it was found that there was a positive relationship between strategic talent management practices and work behaviors of employees. Smits et al. (2014) aimed to reveal the relationship between the strategic talent management practices applied to the personnel in the Radiology department and the success of the personnel. With the research, it was concluded that 80 of the 100 personnel who underwent strategic talent management were successful and the success rate of strategic talent management was 80%. In the study conducted by Ambrosius (2018), the relationship between different strategic talent management practices and employees' intention to leave Brazil's multinational companies was analyzed. It was revealed that linear regression modeling and organizational support and perceived career opportunities were negatively related to the intention of Brazilian employees to leave, while education and development were positively related. The study conducted by Cui et al. (2018) aimed to reveal how strategic talent management is defined and understood by Chinese small and medium-sized enterprises (SMEs). Accordingly, the talent management and retention strategies used by Chinese SMEs in the service sector were examined. It has been concluded that SMEs have different views on strategic talent management. According to some, strategic talent management means having the right candidate in the right business category. The findings also show that a positive work environment, career development opportunities, and a good wage are considered the best strategy to attract talent. In another study, Sheehan et al. (2018) found that consistent strategic talent management practices in the hotel and tourism sectors, especially competitive rewards and training and development opportunities, would improve the brand and have a direct impact on business quality. Another study by Kimathi (2015) examined how strategic talent management affected the performance of Imperial Bank Limited in Kenya. Within the scope of the study, an interview was held with five department managers at Imperial Bank. It has been concluded that the most common strategic talent management practices used by Imperial Bank are performance-based reward system, performance-based promotions, and training programs in terms of annual bonus and salary increases. In another study, Lucie et al. (2016) focused on approaches to the implementation of strategic talent management adopted by agricultural and forestry companies. The results of the study show that 62% of enterprises operating in the agriculture and forestry sector are familiar with the principles of strategic talent management and talent management is part of the mission of the enterprise.

When the results of the studies are examined, it is seen that the enterprises attach importance to the concept of strategic talent. Businesses know the value of talented employees for their own vision and goals and make the necessary investment and staff in this field. In addition, businesses support with features such as salary, promotion, and benefits in order not to lose talented personnel. Although businesses see the concept of strategic talent as an innovative and digital concept, most businesses do not yet seem to have distinguished themselves sufficiently from traditional talent policies. In the study conducted by Lucie et al. (2016), it was found that 62% of the enterprises examined had knowledge about strategic talent management. In this study, it was observed that the rate was close to 100%.

3 A Qualitative Research in the Energy Sector

3.1 Methodology

Within the scope of the research, it is aimed to reveal the perspective and usage areas of strategic talent, which is a new term formed by strategic management and talent management practices that have developed in enterprises in the last 10 years. The preferred method for this purpose is the phenomenological model of the qualitative research method. In this method, the concepts on a certain subject and the details that make up this subject are investigated and in-depth information about the subject is obtained. The research technique frequently preferred in the phenomenology model is interview and observation. In this study, interview technique was used as qualitative research technique. Thanks to the interview technique, detailed information was obtained about the strategic talent applications of the enterprises operating in the energy sector in Turkey. During the evaluation of the data obtained in the research, the NVIVO program was used. With NVIVO, researchers can group the data they obtain in phenomenological research and combine them according to their similar characteristics (Baş & Akturan, 2008). Therefore, the NVIVO program facilitates the evaluation process for qualitative research in terms of providing the opportunity to edit the data.

Companies in the energy sector (https://www.fortuneturkey.com) included in the Fortune 500 research report, which listed the 500 largest companies in Turkey published in 2020, were included in the study. An interview was conducted with the managers and experts of 36 companies in the energy sector. The NVIVO package program was used in the process of grouping the interview data according to similar characteristics and exporting these groups in the form of a graph. Companies participating in the research were given a number between COMPANY1 and COMPANY36 for ease of coding.

Within the scope of the interview, firstly, demographic questions were asked to the participants. In the light of the information obtained from the literature review, the questions prepared for the interview were created in seven dimensions. These dimensions are as follows:

- Questions asked to understand strategic processes in companies (1–5),
- Questions asked to understand the management model in companies (6–10),
- Questions asked to understand backup and career planning activities in companies (11–17),
- Questions asked to understand future planning activities in companies (18–20),
- Questions to understand leadership gaps in companies (21–24),
- Questions asked to understand activities to fill leadership gaps in companies (25–29),
- Questions asked to understand companies' perspective on strategic talent (30–32).

3.2 *Scope of the Study*

When the demographic information of the company representatives participating in the interview is examined, it is seen that 19% of the participants are female and 47% are male. In terms of marital status, 47% of the participants were single, and 42% were married. When the ages of the participants are examined, it is seen that 58% are in the 30–39 age group, 25% in the 24–29 age group, and 14% in the 40–49 age group. The proportion of the participants in the 50–59 age group is 3%. 53% of the participants had bachelor's degree, 31% had a master's degree, and 6% had a doctorate degree. When the total work experience of the participants is examined, it is seen that 33% of the participants have 6–10 years of work experience, 33% have 10 years of work experience and more. While the rate of those with 3–5 years of work experience is 31%, the rate of those with less than 2 years of work experience is 3%. When the titles of the participants are examined, it is seen that 31% are managers, 28% are team leaders, and 14% are senior experts.

4 Conclusion and Discussion

There has been an increase in interest in talent management and strategic talent management over the past decade. Many businesses are convinced that talented staff are useful for their business and have begun to make various moves to attract talented candidates and keep existing talented candidates in business. This has highlighted the concept of strategic talent management. Within the scope of the study, interviews were conducted with the managers of 36 companies operating in the energy sector to understand the perspective of enterprises on strategic talent management. In the interview, the workforce planning, talent planning, perspective on strategic talent management, and the moves made by the enterprises to keep talented employees in the business were evaluated.

When we look at the results of the research, it is seen that evaluations in companies are usually made by meetings or statistical reports. While statistical reports usually generate weekly and monthly controlled reports, meetings are usually held weekly, periodically, and annually. When the time allocated by managers to team career planning is examined, it is seen that some of the timing of career planning is not certain and certain ones are realized monthly. The study conducted by Benoy and Gracias (2015) states that such uncertainties lead to a lack of motivation in the personnel. It is seen that labor force planning is in the form of human resources investments in the majority of enterprises with labor force planning. In other words, it is seen that enterprises attach the necessary importance to labor force planning. When the actions taken by the companies to protect the talents with high potential are examined, it is seen that staff enrichment and training come to the forefront. Considering the studies conducted by Ambrosius (2018), it is seen that the intention of employees to leave can be prevented by training. In this context, it is important for the enterprises examined to invest in education in order to protect their existing talents. When the personnel backup plans of the companies are examined, it is seen that most of the enterprises make backup with training programs. It has been observed that training programs are usually in the form of determining the persons who can act as proxies for the employees. When the studies of the companies for the talents they contain are examined, it is seen that most of the companies work for talents. These works are in the form of staff enrichment such as promotion and raise in most companies. Sripirom et al. (2016) reached a similar conclusion in their study. In this study, a significant relationship was found between personnel enrichment and strategic talent management activities. It is seen that businesses emphasize the training program and small managerial roles while gaining leadership qualifications to their talents. Schreuder and Noorman (2019), on the other hand, stated that to develop these leadership qualities, especially personal development trainings should be focused. However, in this study, no conclusion was reached about the details of the training programs. It is seen that the work done to adapt to the digital workforce of the enterprises in the future is generally to follow current developments and adapt to these developments. Many businesses adapt in terms of digital workforce. When the leadership, functional and technical competencies required by the enterprises to reach their corporate strategies are examined, it is observed that leadership qualification is generally needed. In other words, it is seen that the issues that businesses attach the most importance to within the scope of strategic talent management are leadership. Smits et al. (2014) stated that leadership is seen as important in 80% of the enterprises researched in strategic talent management. Evaluation of the competencies of talented employees of enterprises is usually in the form of periodic evaluations. In other words, there is no comprehensive evaluation. Evaluations are usually carried out through statistical reports. It is seen that most of the enterprises do not have a competence guide. Businesses that do not have a competence guide state that the competence processes proceed improvised and person based. When the ones responsible for the loss of competent personnel in companies are examined, it is generally seen that the responsible persons are managers and human resources units. In addition, it is seen that talent management covers all personnel in enterprises. It is seen that career planning is not only directed at managers in most businesses, but also the personnel other than managers are involved in career planning processes. In the study conducted by Anlesinya et al. (2019), it is stated that strategic talent management should include all individuals instead of a single or a group of individuals. When we examine what resources companies use for staff recruitment processes, it is generally seen that internal and external resources are carried out in a balanced manner. When the policies implemented by companies to keep talents in the company are examined, it is seen that most of them are personnel enrichment such as salary, promotion, and rights. Sheehan et al. (2018) argued that in addition to these, creating a competitive environment would be successful in keeping talents in the business. When the reasons for leaving the job of talented people in enterprises are examined, it is seen that different job offers are the first reason. In other words, the reason behind the resignation of talented personnel is that other companies offer the most talented personnel rights such as better salary and benefits. Cui et al. (2018) argue that a positive working environment and a good wage will increase the likelihood of talented people staying in the current workplace. Therefore, businesses can offer them a good working environment and make improvements in wage rates to maintain their existing capabilities. When the answers given to this question, which reveals the perspectives of enterprises on strategic ability, are examined, it is seen that strategic talent is generally defined as moves towards the company's goal and vision. In other words, strategic talent is considered as the steps taken by most enterprises to achieve the establishment purpose of the enterprise and to obtain a competitive advantage. When the importance of the concept of strategic ability in companies is examined, it is possible to say that the prominent elements are sustainability, success, efficiency, ensuring workflow and accurate evaluation of employees. Many of the participants argue that strategic talent plays an important role in companies achieving a sustainable competitive advantage. Similar results were obtained in the study conducted by Shet (2020). Shet (2020) states that strategic talent management is useful in achieving the company's goals. Similarly, in the study conducted by Karadal (2019), it is stated that strategic talent management contributes to the differentiation of the business from its competitors. On the other hand, in a study conducted by Abubakar and Abdullah (2017), it was stated that strategic talent management would also contribute to businesses based on departments. In this study, no conclusion has been reached on contributing to the departments.

The areas most emphasized by businesses within the scope of strategic talent management were to ensure sustainable competition. According to the results of the research, it is possible to define strategic talent management as activities and processes that include the systematic identification of key positions that contribute differently to the sustainable competitive advantage of the organization, the development and development of a pool of high-potential and high-performance officials to fill these roles.

Businesses that implement strategic talent management systems in accordance with the vision of the business are likely to gain a sustainable competitive advantage. In addition, it is seen that businesses do not show the necessary importance to keep talents. In this context, preparing an appropriate work environment for talents and increasing wages may increase the likelihood of retaining talents. Future research may be on comparing the profitability of businesses that implement and do not implement strategic talent management, or on examining strategic talent management in certain sectors.

The energy sector is an area where competition is very intense. In this context, energy companies need to take some actions to increase their competitiveness (Haiyun et al., 2021; Li et al., 2020). Otherwise, it will be very difficult to survive in such an environment. In this process, one of the important actions is to have qualified personnel. Energy companies do a much more specific business than other industries (Zhao et al., 2021; Zhong et al., 2020). Within this framework, the personnel should also have knowledge about these specific issues. In this process, human resources departments play a very important role. Persons to be employed in the company in line with the needs must have the necessary knowledge (Cheng et al., 2020; Yuan et al., 2021). On the other hand, due to operational processes, the risks faced by energy companies are higher compared to other businesses. Therefore, managers who will work in energy companies must have the necessary skills (Qiu et al., 2020). In this context, energy companies should consider this issue while employing managers (Liu et al., 2021). This situation is especially important for coal and nuclear energy investments. There are significant risks in both types of energy investments. In cases where these risks cannot be managed effectively, accidents with very high damage may occur (Yuan et al., 2020). In this process, a great responsibility falls on the managers of energy companies. In order for these risks to be managed effectively, company managers must take timely action (Du et al., 2020; Yüksel et al., 2020). Therefore, managers in energy companies need to be able to cope with stress much more easily.

References

Abubakar, A., & Abdullah, H. H. (2017). The mediating role of quality culture on the relationship between talent management and organization performance. In *4th conference on business management, 29–30 Oktober 2017*. The Gurney Resprt Hotel & Residences.

Al Ariss, A., Cascio, W. F., & Paauwe, J. (2014). Talent management's: Current theories and future research directions. *Journal of World Business, 49*(2), 173–179.

Ambrosius, J. (2018). Strategic talent management in emerging markets and its impact on employee retention's. Evidence from Brazilian MNCs. *Thunderbird: International Business Review, 60*(1), 53–68.

Angheluţă, S. P., Margina, O., Zaharia, A., & Arionesei, G. (2014). The role of human resources in sustainable development of the energy sector. *Ecoforum Journal, 3*(1), 7–11.

Anlesinya, A., Dartey-Baah, K., & Amponsah-Tawiah, K. (2019). Strategic talent management scholarship: A review of current foci and future directions. *Industrial and Commercial Training, 51*(5), 299–314.

Baş, T., & Akturan, U. (2008). *Nitel araştırma yöntemleri: Nvivo 7.0 ile nitel veri analizi*. Seçkin Yayıncılık.

Benoy, J., & Gracias, N. (2015). Strategic talent management: A proposed model for building talent pipeline of project managers. *International Journal of Management (IJM), 6*(7), 67–83.

Bogomolova, Y. I. (2018). Human capital of transnational corporations in the energy sector. *International Journal of Energy Economics and Policy, 8*(6), 128–134.

Borisova, O. N., Silayeva, A. A., Saburova, L. N., Belokhvostova, N. V., & Sokolova, A. P. (2017). Talent management as an essential element in a corporate personnel development strategy. *Academy of Strategic Management Journal, 1*(16), 31–46.

Chen, S. Y., Lee, A. Y. P., & Ahlstrom, D. (2021). Strategic talent management systems and employee behaviors: The mediating effect of calling. *Asia Pacific Journal of Human Resources, 59*(1), 84–108.

Cheng, F., Lin, M., Yüksel, S., Dinçer, H., & Kalkavan, H. (2020). A hybrid hesitant 2-tuple IVSF decision making approach to analyze PERT-based critical paths of new service development process for renewable energy investment projects. *IEEE Access, 9*, 3947–3969.

Cui, W., Khan, Z., & Tarba, S. Y. (2018). Strategic talent management in service SMEs of China. *Thunderbird International Business Review, 60*(1), 9–20.

Du, L., Dinçer, H., Ersin, İ., & Yüksel, S. (2020). IT2 fuzzy-based multidimensional evaluation of coal energy for sustainable economic development. *Energies, 13*(10). 2453.

Farndale, E., Scullion, H., & Sparrow, P. (2010). The role of the corporate HR function in global talent management. *Journal of World Business, 45*(2), 161–168.

Haiyun, C., Zhixiong, H., Yüksel, S., & Dinçer, H. (2021). Analysis of the innovation strategies for green supply chain management in the energy industry using the QFD-based hybrid interval valued intuitionistic fuzzy decision approach. *Renewable and Sustainable Energy Reviews, 143*, 110844.

Hussaini, I. U., & Majid, N. H. A. (2014). Human behaviour in household energy use and the implications of energy efficiency delivery: A case of Bauchi, Nigeria. *International Journal of Energy Sector Management, 8*(2), 230–239.

Karadal, H. (2019). Stratejik insan kaynakları yönetiminde yeni vizyon: Yetenek yönetimi. In *4th International EMI Entrepreneurship & Social Sciences Congress Proceedings e-book*, 29-–30 November 2019, Istanbul. 1034–1044.

Kimathi, C. M. (2015). *Strategic talent management and performance of imperial bank limited in Kenya*. Doctoral dissertation, University of Nairobi.

Li, X., Zhu, S., Yüksel, S., Dinçer, H., & Ubay, G. G. (2020). Kano-based mapping of innovation strategies for renewable energy alternatives using hybrid interval type-2 fuzzy decision-making approach. *Energy, 211*, 118679.

Liu, Y., Gong, X., Yüksel, S., Dinçer, H., & Aydın, R. (2021). A multidimensional outlook to energy investments for the countries with continental shelf in East Mediterranean region with hybrid decision making model based on IVIF logic. *Energy Reports, 7*, 158–173.

Lucie, V., Hana, U., & Helena, S. (2016). Strategic talent management in agricultural and forestry companies. *Agricultural Economics, 62*(8), 345–343.

Markovska, N., Taseska, V., & Pop-Jordanov, J. (2009). SWOT analyses of the national energy sector for sustainable energy development. *Energy, 34*(6), 752–756.

Muda, I., Wardani, D. Y., Maksum, A., Lubis, A. F., Bukit, R., & Abubakar, E. (2017). The influence of human resources competency and the use of information technology on the quality of local government financial report with regional accounting system as an intervening. *Journal of Theoretical & Applied Information Technology, 95*(20), 5552–5561.

Qi, W., Huang, Z., Dinçer, H., Korsakienė, R., & Yüksel, S. (2020). Corporate governance-based strategic approach to sustainability in energy industry of emerging economies with a novel interval-valued intuitionistic fuzzy hybrid decision making model. *Sustainability, 12*(8), 3307.

Qiu, D., Dinçer, H., Yüksel, S., & Ubay, G. G. (2020). Multi-faceted analysis of systematic risk-based wind energy investment decisions in E7 economies using modified hybrid modeling with IT2 fuzzy sets. *Energies, 13*(6), 1423.

Schreuder, R., & Noorman, S. (2019). Strategic talent management: Creating strategic value by placing top talents in key positions. *Development and Learning in Organizations: An International Journal, 33*(1), 1–4.

Sheehan, M., Grant, K., & Garavan, T. (2018). Strategic talent management, s. a macro and micro analysis of current issues in hospitality and tourism. *Worldwide Hospitality and Tourism Themes, 10*(1), 28–41.

Shet, S. V. (2020). Strategic talent management–contemporary issues in international context. *Human Resource Development International, 23*(1), 98–102.

Smits, C., Bertelsen, R. G., & Justinussen, J. C. S. (2014). The challenges & opportunities for Arctic microstates in developing an energy sector: The role of human capital and knowledge institutes. *Arctic Yearbook, 3*(1), 1–17.

Sparrow, P. R., & Makram, H. (2015). What is the value of talent management? Building value-driven processes within a talent management architecture. *Human Resource Management Review, 25*(3), 249–263.

Sripirom, K., Jhundra-indra, P., & Raksong, S. (2016). Strategic talent management and firm success: Evidence from electronic and electrical appliance businesses in Thailand. *AU-GSB e-Journal, 9*(2), 31–31.

Yuan, G., Xie, F., Dinçer, H., & Yüksel, S. (2021). The theory of inventive problem solving (TRIZ)-based strategic mapping of green nuclear energy investments with spherical fuzzy group decision-making approach. *International Journal of Energy Research, 45*(4), 6435.

Yuan, J., Zhang, Z. M., Yüksel, S., & Dinçer, H. (2020). Evaluating recognitive balanced scorecard-based quality improvement strategies of energy investments with the integrated hesitant 2-tuple interval-valued Pythagorean fuzzy decision-making approach to QFD. *IEEE Access, 8*, 171112–171128.

Yüksel, S., Dinçer, H., & Uluer, G. S. (2020). The role of technological development on renewable energy usage: An econometric analysis for G7 countries. In *Handbook of research on sustainable supply chain management for the global economy* (pp. 136–153). IGI Global.

Zhao, Y., Xu, Y., Yüksel, S., Dinçer, H., & Ubay, G. G. (2021). Hybrid IT2 fuzzy modelling with alpha cuts for hydrogen energy investments. *International Journal of Hydrogen Energy, 46*(13), 8835–8851.

Zhong, J., Hu, X., Yüksel, S., Dinçer, H., & Ubay, G. G. (2020). Analyzing the investments strategies for renewable energies based on multi-criteria decision model. *IEEE Access, 8*, 118818–118840.

Relationships between Energy Efficiency on Output and Energy Efficiency on Carbon Emission

Imran Hussain, Swarup Samanta, and Ramesh Chandra Das

1 Introduction

With the progress of globalization across the world, the countries from the developing zones are now increasingly dependent upon using their energy base to produce more and contribute more to their gross domestic products. There are so many channels such as industry and service activities through which energy use helps in production activities. But there are several negative aspects to these energy uses. Countries generate several pollutants and emit to the ambient environment which has huge health cost due to environmental damage. Further, there are huge unnecessary uses of energy as it does not contribute to the country's output at its potential. Therefore, there is a natural question on whether energy use has any sort of positive effects upon the outputs of the economies from developing zones. This is particularly important for developing economies of South-Asia region where per capita GDP of this region [PPP (current international $ 4934.3)] is still much lower than that of the low- and middle-income countries [PPP (current international $ 9211.8)] (World Bank, 2014). However, it is exciting that the South-Asian nations are growing well in recent years. In 2014, India experienced 7.4% annual GDP growth rate ranking the country first, followed by Bangladesh with 6.1% GDP growth rate ranking the country second, Nepal with 6.0% growth rate ranking the country third, and Sri Lanka with 5.0% growth rate ranking the country fourth in the region. Pakistan experienced the lowest GDP growth rate in the region which was 4.7% (World Bank, 2014). In spite of other important factors, economic growth is highly affected by energy use. The data of World Bank says that the highly developing countries are

I. Hussain · S. Samanta
Research Scholar, Department of Economics, Vidyasagar University, Midnapore, India

R. C. Das (✉)
Department of Economics, Vidyasagar University, Midnapore, West Bengal, India

© The Author(s), under exclusive license to Springer Nature Switzerland AG 2022
H. Dinçer, S. Yüksel (eds.), *Sustainability in Energy Business and Finance*, Contributions to Finance and Accounting,
https://doi.org/10.1007/978-3-030-94051-5_7

now chasing the so-called developed countries in terms of energy use and atmospheric pollution. Literature suggests that energy use which results in CO_2 emissions play controversial roles towards economic growth [Cheng (1999), Chang (2010), Rahman et al. (2020), Lu (2017), Vo et al. (2019), etc.]. The present study, under this backdrop, aims to examine the long-run associations of 'national output to energy use' and 'carbon emission to energy use' in five selected South-Asian nations: India, Pakistan, Bangladesh, Sri Lanka, and Nepal for the period of 1971–2014 on the World Bank data.

2 Review of the Existing Literature

Several studies on the relationship among the three variables are available from the literature. Relevant studies are briefly discussed below for the present study. A recent investigation based on extended neoclassical economic growth model was done by Rahman et al. (2020) to expose the interconnection of carbon emissions, population density, and trade openness on economic growth for five South Asian countries for the period of 1990–2017 and shows that CO_2 emissions and population density positively and significantly affect the economic growth permanently in South Asia while trade openness affects economic growth negatively. There is a bi-directional long-run causality between CO_2 emissions and economic growth and unidirectional causal link running from population density to CO_2 emissions. Finally, there is a unidirectional causal link running from labour to economic growth and population density.

Li et al. (2016) calculated the degree of association among energy consumption, environmental pollution, and economic growth by applying gray correlation analysis for China. The study found that energy consumption motivates development of the economy and causes environmental pollution. According to Chang (2010), economic growth induces a higher level of energy consumption and CO_2 emissions. The results demonstrate bi-directional causality running from GDP to CO_2 emissions and secondly, electricity consumption to GDP. Another work on China by Zeng et al. (2020) studied on the energy consumption, FDI, and development in Zhejiang, for the period of 1993–2017. Long-run relationship exhibits among these variables. In terms of short-term dynamic relationship, there was a strong positive relationship between FDI and energy consumption. The result was consistent with the findings of Zhao et al. (2007) and Latif et al. (2021). In contrast of this, energy consumption played a strong positive impact on economic growth and is consistent. After China's accession to the WTO, a higher level of opening boosted the energy. The investigation of Wang et al. (2016) for the same country shows the existence of cointegrating relationship among economic growth, energy consumption, and CO_2 during the period of 1990–2012. A bi-directional causal relationship between economic growth and energy consumption was emerged.

Yu and Hwang (1984) postulate the causality between GNP and energy consumption for US by using time series data for the period of 1947–1979. The result shows that there is no causal relationship between GNP and energy consumption. The further result also exhibits of no causal relationship between per capita energy and per capita GNP. The dynamic causal relationships between pollutant emissions, energy consumption, and output are applied by Ang (2007) for France during the period of 1960–2000. Using cointegration and vector error correction model, it provided a result that variables are strongly inter-related and therefore their relationship must be examined using an integrated framework. Similarly, Halicioglu (2009) attempts to empirically examine the dynamic causal relationships among carbon emissions, energy consumption, income, and foreign trade in Turkey for the period of 1960–2005. For Iran, Taghavee et al. (2016) find out various socioeconomic elasticities in the long run and short run during 1974–2012. The results based on per capita CO_2 emissions, GDP, and energy consumption proved the strong elasticities in the long run. Trade openness, labour force, and financial development play the most leading role in the short run, even though their limited role in the long run.

Lee (2006) uses the causality testing procedure to examine the causal interplays between energy consumption and GDP in the G-11 countries. Neutral causality is observed in United Kingdom, Germany, and Sweden. Bi-directional causality exists in the United States but unidirectional causality running from energy consumption to GDP in Canada, Belgium, the Netherlands, and Switzerland. Similar study has been done by Narayan and Smyth (2008) on G7 countries to examine the relationship between capital formation, energy consumption, and real GDP using panel econometrical methods. The result was that capital formation, energy consumption, and real GDP are cointegrated and real GDP is positively influenced by capital formation and energy consumption.

Akbostancı et al. (2009) arranged two types of data for 58 provinces of Turkey. In the first stage, time series model covers 1968–2003, and the panel data model covers 1992–2001 in the second stage. According to time series analysis, a monotonically increasing relationship between CO_2 and income exists in the long run. But the panel data analysis shows an N-shaped relationship for SO_2 and PM10 emissions that does not support the Environmental Kuznets Curve hypothesis. For the same country, Ozturk and Acaravci (2013) investigated the causality between financial development, openness, economic growth, energy consumption, and carbon emissions in for 1960–2007. Environment Kuznets curve hypothesis was validated in this economy. On the contrary, short-run unidirectional causality exists from financial development to per capita energy consumption and per capita real income. The similar hypothesis was applied by Rahman (2017) to expose the relation between CO_2 emissions, energy use, economic growth, exports, and population density for a panel of 11 Asian populous countries over the period of 1960–2014. The existence of U-shaped relationship between real income and CO_2 emissions in the panel of 11 countries but the inverted U-shaped relationship is found only for the Philippines. Furthermore, environmental quality is adversely affected by energy use, exports, and population density in the panel of these countries in the long run. Another research was done by Saboori et al. (2012) for Malaysia on Environmental Kuznets Curve hypothesis. The result shows the existence of long-run relationship between per capita carbon emissions and real per capita income. Saidi and Hammami (2015) examined the effect of economic growth and CO_2 emissions on energy consumption using a growth framework and simultaneous equation models. The results show that the effect of economic growth on energy use is positive and statistically significant in the global panel of 58 countries. Dinda et al. (2000) used panel data set of 88 countries over the period of 1960–1990 and found a long-run relationship between per capita income and per capita carbon emission. The obtained results also suggest that there exists a bi-directional causal relationship between these two variables for Africa, Central America, America as a whole, Eastern Europe, Western Europe, Europe as a whole, and the World as a whole.

Alkhathlan and Javid (2013) revealed a relationship among economic growth, carbon emissions, and energy consumption in Saudi Arabia both in the aggregate and disaggregate levels. Both the long-run and short-run income elasticities of carbon emissions are negative for the gas consumption model. Using the same variables, Ang (2008) shows the long-run association between output, energy, and emissions in Malaysia during the period of 1971–1999. The observation was that pollution and energy use are positively related to output in the long run. The causality result shows that energy consumption growth is caused by economic growth, both in the short run and long run. Similarly, Kais and Ben Mbarek (2015) selected three North African countries—Algeria, Egypt, and Tunisia for the period of 1980–2012. From the cointegration test results, they found a long-run relationship between CO_2 emission and energy of the three countries.

Hossain (2012) explores the dynamic causal relationship between carbon emissions, energy consumption, economic growth, foreign trade, and urbanization for the period of 1960–2009 in Japan. In the short run, carbon emission is influenced by both the energy consumption and trade openness, economic growth is affected by carbon emissions, and trade openness is influenced by economic growth. It is also found that in the long run, higher energy consumption leads to more carbon emissions. Using modified version of Granger causality test, Menyah and Wolde-Rufael (2010) conclude that nuclear energy consumption leads to increase in carbon emissions without feedback in US during the period of 1960–2007. They show econometrical evidence and suggest that nuclear consumption can help to reduce carbon emissions, but so far, the renewable energy consumption has not reached a level where it makes a significant contribution to emissions reduction. Rahman and Kashem (2017) applied the ARDL Bounds Testing methodology in the case of Bangladesh over the period of 1972–2011. Both the industrial production and energy consumption significantly and positively impact the carbon emissions in the short as well as long runs. The result shows the existence of a unidirectional causality running from industrial development to CO_2 emissions, and industrial development to energy consumption and energy consumption to CO_2 emissions.

3 Rationale of the Study

From the brief review of the existing literature, it is revealed that most of the research works show the relationship of economic growth, energy used, and carbon emissions separately. Since both the national income and carbon emissions simultaneously depend on energy consumption, therefore, in this study, ratio form is taken to analyse the proper relationship between 'national output to energy used' and 'carbon emission to energy used'. The first one represents average productivity of energy upon GDP and the second one is average productivity of energy upon carbon emissions.

The present research study has three key variables: Gross Domestic Product (GDP) as measured in current US$, Carbon emissions ($CO_2$) as measured in kt., and the last variable is Energy used (kg of oil equivalent per capita). To show the average productivity with respect to energy used, the present study takes ratio form such as National Output to Energy Used (NOEU) and Carbon Emission to Energy Used (CEEU). CO_2 emissions are those stemming from the burning of fossil fuels and the manufacture of cement. They include carbon dioxide produced during consumption of solid, liquid, and gas fuels and gas flaring. Energy use refers to use of primary energy before transformation to other end-use fuels, which is equal to indigenous production plus imports and stock changes, minus exports and fuels supplied to ships and aircraft engaged in international transport.

The data from 1971 to 2014 regarding Energy used, GDP, and CO_2 are used in the study is secondary annual data and is collected from World Bank (https://data.worldbank.org). The present study is based on the hypotheses of testing the cointegration and causality between average productivity of energy used upon GDP (NOEU) and average productivity of energy used upon CO_2 (CEEU) for different South-Asian nations.

Initially, Pearson's correlation coefficient is used to investigate the degree of association between NOEU and CEEU. Then time series econometric method is used chronologically to verify the cointegration and causality between these two variables. Augmented Dickey and Fuller (1979) test including m lags of the dependent variable to correct any serial correlation in the disturbance term is used for test of stationarity. The present research study selects the equation with intercept from the three augmented models, which can be estimated and the null hypothesis H_0: $\delta = 0$ can be tested by using a τ-statistic.

ADF test for NOEU:

$$\Delta(\text{NOEO})_t = \alpha + \delta(\text{NOEU})_{t\text{-}1} + \sum\nolimits_{i=1}^{m} \gamma_i \Delta(\text{NOEU})_{t\text{-}i} + u_t \quad (1)$$

ADF test for CEEU:

$$\Delta(\text{CEEU})_t = \alpha + \delta(\text{CEEU})_{t\text{-}1} + \sum\nolimits_{i=1}^{m} \gamma_i \Delta(\text{CEEU})_{t\text{-}i} + u_t \quad (2)$$

Where u_t is a white noise error term. These equations incorporate difference terms $\Delta(\text{NOEU})_{t-1} = (\text{NOEU}_{t-1} - \text{NOEU}_{t-2})$, $\Delta(\text{NOEU})_{t-2} = (\text{NOEU}_{t-2} - \text{NOEU}_{t-3})$, etc., the similar process is applicable for Eq. (2).

If the computed absolute value of the tau statistics (τ) exceeds the ADF or Mackinnon critical values, we reject the null hypothesis that H_0: $\delta = 0$, in which case the time series is stationary. The ADF test is based on the assumptions that the error term is serially independent and has a constant variance. Thus, Phillips and Perron developed a generalization of the ADF test procedure where no such restrictive assumptions on the distribution of error terms are there. The test regression equation is same as mentioned in (Eqs. 1 and 2) but the PP test makes a correction to the computed τ-statistic of the estimated coefficient of δ to account for serial correlation in u_t. Further, it has been shown that asymptotic distribution of the PP τ-statistic is the same as the distribution of ADF τ-statistic. So, the ADF critical values are still applicable here.

4 Methodology

Once confirmation of stationarity of the data series is obtained, the next step is to check up whether the series are integrated or not, i.e. whether there exists of long-run relationship between NOEU and CEEU. Engle and Granger (1987) provided a test to examine the presence of cointegrating relationship (i.e. long-run relationship) between the variables through the residual's time series behaviour. The method runs through the following steps:

Step I: If both variables are $I(0)$, it is not necessary to proceed to test the possibility of cointegration. Here one should continue with the OLS regression analysis since standard time series methods are applicable to the stationary variables. If the variables are integrated of different order, they are not cointegrated. If both the variables are $I(1)$, apply the OLS method to estimate equation [say, $\text{NOEU}_t = \alpha + b(\text{-CEEU}_t) + \varepsilon_t$] and generate the series of estimated residuals $\widehat{\varepsilon}_t = \text{NOEU}_t - \widehat{a} - \widehat{b}(\text{CEEU}_t)$.

Step II: For checking the stationarity of $\widehat{\varepsilon}_t$, ADF test is applied. The form of the ADF test equation here is

$$\Delta\widehat{\varepsilon}_t = \delta\widehat{\varepsilon}_{t-1} + \sum\nolimits_{i=1}^{m} \alpha_i \Delta\widehat{\varepsilon}_{t-1} + v_t \tag{3}$$

Now testing the null hypothesis H_0: $\delta = 0$, we conclude whether $\widehat{\varepsilon}_t$ is $I(0)$ (i.e. stationary) or not. When the H_0 gets rejected, our conclusion is that $\widehat{\varepsilon}_t \tilde{I}(0)$ and the variables are cointegrated.

The cointegrating equation gives long-run relationships between the two variables. However, cointegrating equation does not shed any light on short-run dynamics although its existence indicates that there must be some short-term forces that are responsible for keeping the long-run relationship intact. This entails the construction of error correction mechanism to model dynamic relationship. The Purpose of the Error Correction Model is to indicate the speed of adjustment from the short run equilibrium to the long-run equilibrium state. The equation is

$$\Delta(\mathrm{NOEU})_t = \phi + \gamma\Delta(\mathrm{CEEU})_t + \lambda\widehat{\varepsilon}_{t-1} + \eta_t \qquad (4)$$

Where γ = Short-run coefficient, (measure the immediate impact of a change in CEEU_t will have on a change in NOEU_t, λ = error correction coefficient and shows how much of the disequilibrium is being corrected, $\widehat{\varepsilon}_{t-1}$ = error correction term. Here $\widehat{\varepsilon}_{t-1} = \mathrm{NOEU}_{t-1} - \widehat{\mathrm{NOEU}}_{t-1}$ is one period lagged value of the error from the cointegrating regression' η_t = white noise error term in the ECM. When $\widehat{\varepsilon}_{t-1}$ is non-zero (positive or negative), there is disequilibrium in the short run. However, equilibrium will be restored in the long run if and only if $\lambda < 0$.

Granger (1969) was the first econometrician to offer a formal test of the direction of causality between the variables. It is basically a statistical test. The Granger test involves estimating the following pairs of equations:

$$\mathrm{NOEU}_t = a_1 + \sum_{i=1}^{n} \alpha_i \mathrm{CEEU}_{t-i} + \sum_{j=1}^{m} \beta_j (\mathrm{NOEU})_{t-j} + \varepsilon_{1t} \qquad (5)$$

$$\mathrm{CEEU}_t = a_2 + \sum_{i=1}^{n} \gamma_i (\mathrm{CEEU})_{t-i} + \sum_{j=1}^{m} \delta_j (\mathrm{NOEU})_{t-j} + \varepsilon_{2t} \qquad (6)$$

Where ε_{1t} and ε_{2t} are uncorrelated white noise error terms. The null hypotheses are

$H_0 := \alpha_i = 0 \ (i = 1,\ 2, \ldots,\ n)$ [For Eq. 5].
$H_0 := \delta_j = 0 \ (j = 1,\ 2, \ldots,\ m)$ [For Eq. 6].
The test statistic

$$F^* = \frac{(\mathrm{RSS}_{\mathrm{restricted}} - \mathrm{RSS}_{\mathrm{unrestricted}})/m}{\mathrm{RSS}_{\mathrm{unrestricted}}/(n-k)} \qquad (7)$$

If the computed-*F* exceeds the critical-*F* value [i.e. $F^* > F_\lambda\,(m, n\text{-}k)$], then reject the H_0 and conclude that 'CEEU Granger causes NOEU' (for Eq. 5) or 'NOEU Granger causes CEEU' (for Eq. 6).

It is to note that if the variables are stationary in their first difference, we then apply this test on the first differenced variables in (Eqs. 5 and 6).

5 Empirical Results and Discussion

Before going into the econometric exercises, the study computes the degree of linear associations between the two variables, NOEU and CEEU, for all the selected countries. Then the econometric exercises follow. The correlation coefficient

Table 1 Correlation coefficient between NOEU and CEEU

Country	1971 to 2014		
	r	t-stat	Probability
India	0.9094	14.1699	0.0000
Pakistan	0.9405	17.9377	0.0000
Bangladesh	0.9021	13.5478	0.0000
Sri Lanka	0.8694	11.4030	0.0000
Nepal	0.9296	16.3456	0.0000

Note: r: Correlation Coefficient between NOEU and CEEU
Source: Authors' calculations

between NOEU and CEEU for all the South-Asian nations under study is presented in Table 1 to see the degree of associations between the two across the countries in the list.

From Table 1, it is clearly observed that the average productivity of energy upon GDP and average productivity of energy upon CO_2 simultaneously increased during the period of 1971–2014. As the energy is used in production process, GDP increases and at the same time carbon emission also increases. Nowadays, it is a crucial issue regarding the sustainable development for the nations. In this case technological advancement plays an important role. This is because advancement of technology leads to increment in productivity with reduction in use of energy. It is to note that correlation results do not depict any sort of causal relations between the two variables. Hence, the following econometric exercises are done. The unit root test result (following ADF and PP test) in first difference and second difference of series for South-Asian nations are given in Table 2. Here the null hypothesis, H_0 = Series has a unit root (Non-stationary series).

Series are non-stationary at levels in all the countries. The results conclude that the series are in first difference/second difference form doesn't have a unit root and are stationary. It is found that the hypotheses of unit root for the series are rejected under PP Test also. For India and Bangladesh, series are $I(1)$. The cointegration technique is applied for these two nations to justify the long-run relationship between average productivity of energy used upon GDP and the average productivity of energy used upon carbon emissions. Pakistan, Sri Lanka, and Nepal have no unit root in second difference, i.e. the series are $I(2)$. The validation of long-run relationship and short-run interplays for India and Bangladesh is shown in the following section.

As mentioned in methodology, Engle and Granger (1987) make available a test to examine the presence of cointegrating relationship between the variables. If both the NOEU and CEEU are $I(1)$ and the estimated residuals are $I(0)$, then we may conclude that there exists a long-run relation between them. Thus, ADF test is conducted for India and Bangladesh to show whether the estimated residuals are stationary in level or not.

The result shows that the residual series is stationary in level (the null hypothesis of a unit root is rejected) for Bangladesh only. This implies that both the variables NOEU and CEEU are cointegrated, i.e. the average productivity of energy used upon

Table 2 Augmented Dickey–Fuller (ADF) and Phillips–Perron (PP) Test (1971–2014)

Country	With intercept		
India	ADF	PP	Remarks
ΔNOEU	−5.5215 (0.0000)	−5.5294 (0.0000)	Stationary in first difference
ΔCEEU	−6.0633 (0.0000)	−6.7024 (0.0000)	
Pakistan			
ΔΔNOEU	−5.8814 (0.0000)	−35.767 (0.0001)	Stationary in second difference
ΔΔCEEU	−10.171 (0.0000)	−48.383 (0.0001)	
Bangladesh			
ΔNOEU	−5.9058 (0.0000)	−5.9093 (0.0000)	Stationary in first difference
ΔCEEU	−41.044 (0.0001)	−40.688 (0.0001)	
Sri Lanka			
ΔΔNOEU	−13.033 (0.0000)	−19.501 (0.0001)	Stationary in second difference
ΔΔCEEU	−6.0908 (0.0000)	−32.579 (0.0001)	
Nepal			
ΔΔNOEU	−6.1128 (0.0000)	−12.366 (0.0000)	Stationary in second difference
ΔΔCEEU	−7.4938 (0.0000)	−29.995 (0.0001)	

Note: tau statistics, τ-value is denoted by minus figured value and P-value is in first bracket, 'Δ' and 'ΔΔ' the series are in first difference and second difference, respectively
Source: Authors' own calculations

GDP and average productivity of energy used upon carbon emissions have a long-lasting relationship between them. The variables are not cointegrated (no long-run relationship between them) for India. This alternatively means that energy use is making co-impacts upon both the national output and carbon emission in Bangladesh, not in India. So, Bangladesh is considered for ECM estimation while India is exempted to proceed for ECM estimation for the short run dynamics between NOEU and CEEU. Since cointegration has become possible for Bangladesh, the estimated Error Correction Model for this nation is

$$\Delta\widehat{\text{NOEU}} = 14.02463 + 0.445608\ \Delta\text{CEEU} - 0.150459\ (\text{res})_{t-1}$$
$$p:\quad [0.0050][0.1089][0.2991]$$

The coefficient of $(\text{res})_{t-1}$ is desirably negative but insignificant. The negative coefficient of $(\text{res})_{t-1}$ implies the correction of short-term disturbance to the long-run stable relationship between average productivity of energy used upon GDP and the average productivity of energy used upon carbon emissions. The coefficient of $(\text{res})_{t-1}$ is −0.1504 which shows the speed of adjustment towards equilibrium. Here the speed of adjustment is 15.04% annually. The result of ECM further concludes that there is no presence of long-run causality from CEEU to NOEU. This implies average productivity of energy upon GDP is not influenced instantly by average productivity of energy upon carbon emissions in Bangladesh in the long run. This is our final stage of investigation to show the causal interplays between NOEU and CEEU for the countries in the short run. All the countries are taken for the analysis,

Table 3 Granger Causality test results for sample 1971 to 2014 by nations of South-Asia

Country	Null Hypotheses	Lag	F-stat	Prob.	Conclusion
India	ΔNOEU ↛ ΔCEEU	2	5.2398	0.0101	NOEU → CEEU
	ΔCEEU ↛ ΔNOEU	2	4.6321	0.0162	CEEU → NOEU
Pakistan	ΔΔNOEU ↛ ΔΔCEEU	2	3.7057	0.0347	NOEU → CEEU
	ΔΔCEEU ↛ ΔΔNOEU	2	2.9374	0.0662	CEEU →NOEU
Bangladesh*	ΔNOEU ↛ ΔCEEU	4	1.1854	0.3373	No causality
	ΔCEEU ↛ ΔNOEU	4	6.3173	0.0008	CEEU → NOEU
Sri Lanka	ΔΔNOEU ↛ ΔΔCEEU	2	1.8780	0.1680	No causality
	ΔΔCEEU ↛ ΔΔNOEU	2	0.2599	0.7726	No causality
Nepal	ΔΔNOEU ↛ ΔΔCEEU	3	1.8816	0.1525	No causality
	ΔΔCEEU ↛ ΔΔNOEU	3	3.4622	0.0276	CEEU → NOEU

Note: '*' implies cointegrated nation, '↛' implies 'does not Granger cause', '→' denotes the direction of causality, Lag order selected by majority result of AIC, SIC, and HQ
Source: Author's calculations

but the exercise and interpretations of the results depend upon the orders of integrations of the two series across the countries. The results are presented in Table 3.

The results clearly show the existence of bi-directional causality between NOEU and CEEU in India and Pakistan. This implies variables are mutually dependent to each other, i.e. average productivity of energy upon GDP is influenced by the average productivity of energy used upon carbon emission and at the same time average productivity of energy used upon carbon emission is influenced by the average productivity of energy upon GDP. In other words, when GDP increases due to use of energy, carbon emissions also increase and at the same period, energy leads to carbon emissions which also lead to increase in GDP for these two nations. Increase in income leads to increase in investment in productive sector and that directs the use of energy which show the way of carbon emissions. There are two nations, Bangladesh and Nepal, which have a significant indication of unidirectional causal relationship running from CEEU to NOEU. In case of Bangladesh, carbon emission per unit of energy consistently affects the income per unit of energy used but the converse causality did not happen during this period. The increment of GDP is due to the increment of CO_2 via increase in use of energy. Here the carbon emission is the result of negative externality in the production process. Finally, there is no causal relationship between these two variables in Sri Lanka. GDP per unit use of energy neither is affected nor affect the carbon emission per unit use of energy in this country.

6 Conclusions

The study investigated the existence of long-run association between the efficiency of energy towards the national output (NOEU) and the efficiency of the energy use to environmental pollution (CEEU) using appropriate time series econometric tools.

The cointegration result revealed the existence of long-run associations between these two variables in Bangladesh only. Further, employing the Granger causality test to see the short-run interplays, bi-directional relationship between NOEU and CEEU has been found in India and Pakistan. There are two nations, Bangladesh and Nepal, which have a significant indication of unidirectional causal interplays from CEEU to NOEU. Finally, there is no causal relationship between these two variables in Sri Lanka.

The causal relations from energy use to carbon emission with that of national output are the source of instability in the environmental stock of inputs. The sustainable developmental goals of the nations may get hurt by this causal interplays. The policy makers should think of this issue and frame policies for alternative energy uses.

References

Akbostancı, E., Türüt-Aşık, S., & Tunç, G. P. (2009). The relationship between income and environment in Turkey: Is there an environmental Kuznets curve? *Energy Policy, 37*(3), 861–867.

Alkhathlan, K., & Javid, M. (2013). Energy consumption, carbon emissions and economic growth in Saudi Arabia: An aggregate and disaggregate analysis. *Energy Policy, 62*, 1525–1532.

Ang, J. B. (2007). CO_2 emissions, energy consumption, and output in France. *Energy Policy, 35*, 4772–4778.

Ang, J. B. (2008). Economic development, pollutant emissions and energy consumption in Malaysia. *Journal of Policy Modeling 30*(2), 271–278.

Chang, C. C. (2010). A multivariate causality test of carbon dioxide emissions, energy consumption and economic growth in China. *Applied Energy, 87*, 3533–3537.

Cheng, B. S. (1999). Causality between energy consumption and economic growth in India: an application of co-integration and error-correction modelling. *Indian Economic Review, 1*, 39–49.

Dickey, D., & Fuller, W. (1979). Distribution of the Estimator for Autoregressive Time series with a Unit Root. *Journal of the American Statistical Association, 74*, 427–431.

Dinda, S., Condoo, D., & Pal, M. (2000). Air quality and economic growth: An empirical study. *Ecological Economics, 34*, 409–423.

Engle, R., & Granger, C. (1987). Co-integration and error correction: Representation, estimation, and testing. *Econometrica, 55*(2), 251–276.

Granger, C. (1969). Investigating Causal Relations by Econometric Models and Cross-spectral Methods. *Econometrica, 37*(3), 424–438.

Halicioglu, F. (2009). An econometric study of CO_2 emissions, energy consumption, income and foreign trade in Turkey. *Energy Policy, 37*, 1156–1164.

Hossain, S. (2012). An econometric analysis for CO_2 emissions, energy consumption, economic growth, foreign trade and urbanization of Japan. *Low Carbon Economy, 03*(03), 92–105.

Kais, S., & Ben Mbarek, M. (2015). Dynamic relationship between CO_2 emissions, energy consumption and economic growth in three North African countries. *International Journal of Sustainable Energy, 36*(9), 840–854.

Latif, K., Javeed, S. A., & Sattar, U. (2021). Carbon emission in the SAARC countries with causal effects of FDI, Economic growth and other economic factors. *International Journal of Environmental Research and Public Health., 18*(9), 4605.

Lee, C. C. (2006). The causality relationship between energy consumption and GDP in G-11 countries revisited. *Energy Policy, 34*(9), 1086–1093.
Li, G., Liu, W. L., Wang, Z. H., & Liu, M. Q. (2016). An empirical examination of energy consumption, behavioral intention, and situational factors: Evidence from Beijing. *Annals of Operations Research, 255*(1–2), 507–524.
Lu, W. C. (2017). Greenhouse gas emission, Energy consumption and economic growth: A panel cointegration analysis for 16 Asian countries. *International Journal of Environmental Research and Public Health, 14*(11).
Menyah, K., & Wolde-Rufael, Y. (2010). CO_2 emissions, nuclear energy, renewable energy and economic growth in the US. *Energy Policy, 38*(6), 2911–2915.
Narayan, P. K., & Smyth, R. (2008). Energy consumption and real GDP in G7 countries: New evidence from panel cointegration with structural breaks. *Energy Economics, 30*(5), 2331–2341.
Ozturk, I., & Acaravci, A. (2013). The long-run and causal analysis of energy, growth, openness and financial development on carbon emissions in Turkey. *Energy Economics, 36*, 262–267.
Rahman, M. M. (2017). Do population density, economic growth, energy use and exports adversely affect environmental quality in Asian populous countries? *Renewable and Sustainable Energy Reviews, 77*, 506–514.
Rahman, M. M., & Kashem, M. A. (2017). Carbon emissions, energy consumption and industrial growth in Bangladesh: Empirical evidence from ARDL cointegration and Granger causality analysis. *Energy Policy, 110*, 600–608.
Rahman, M. M., Saidi, K., & Mbarek, M. B. (2020). Economic growth in South Asia: the role of CO_2 emissions, population density and trade openness. *Heliyon, 6*(5), e03903.
Saboori, B., Sulaiman, J., & Mohd, S. (2012). Economic growth and CO_2 emissions in Malaysia: A cointegration analysis of the Environmental Kuznets Curve. *Energy Policy, 51*, 184–191.
Saidi, K., & Hammami, S. (2015). The impact of CO_2 emissions and economic growth on energy consumption in 58 countries. *Energy Reports, 1*, 62–70.
Taghavee, V., Aloo, A., & Shirazi, J. (2016). Energy, environment, and economy interactions in Iran with cointegrated and ECM simultaneous model. *Procedia Economics and Finance, 36*, 414–424.
Vo, A. T., Vo, D. H., Le, Q., & T-T. (2019). CO_2 emission and, energy consumption and economic growth: New evidence in the ASEAN countries. *Journal of Risk and Financial Management, 12*, 145.
Wang, S., Li, Q., Fang, C., & Zhou, C. (2016). The relationship between economic growth, energy consumption, and CO_2 emissions: Empirical evidence from China. *Science of the Total Environment, 542*, 360–371.
Yu, E. S. H., & Hwang, B.-K. (1984). The relationship between energy and GNP: Further results. *Energy Economics, 6*(3), 186–190.
Zeng, S., Liu, Y., Ding, J., & Xu, D. (2020). An empirical analysis of energy consumption, FDI and high quality development based on time series data of Zhejiang province. *International Journal of Environmental Research and Public Health, 19*(9), 3321.
Zhao, X., Hu, J., & Shi, X. (2007). An empirical analysis of the influence of industrial FDI distribution on Chinese energy consumption. *Finance and Trade Economics, 3*, 117–120.

Examination of the Relationship between Economic Growth, Natural Resources, Energy Consumption, Urbanization, and Capital

Mahmut Sami Duran and Şeyma Bozkaya

1 Introduction

One of the ultimate and fundamental goals of economies is to ensure economic growth. It is very important to determine the factors that determine the dynamics of economic growth and to focus on these factors in terms of determining the policies to be implemented. A general consensus has emerged that the main factors that determine the growth trend and continuity of the country's economies are the savings tendency of the society (capital accumulation), invention flows and innovation (the productivity determines the growth rate) and population growth and that these factors should be emphasized (Kaldor, 1957). In this context, studies investigating the basic dynamics and dimensions of economic growth find a wide field of study in the literature (Gylfason, 2001; Cavalcanti et al., 2011; Bal et al., 2016; Baz et al., 2019; Awodumi & Adewuyi, 2020). These studies examining the basic dynamics of economic growth in the literature differ in terms of the variables they use. Growth theories generally accept that the effective input factors of growth are labor, capital, natural resources, the quality of human capital and especially the existence of technological progress.

In addition to these variables, which growth theories accept as basic, factors such as energy use and efficiency, trade openness, human capital, and urbanization population are also considered to be quite effective in explaining economic growth.

M. S. Duran
Department of Finance, Banking and Insurance, Selcuk University, Konya, Turkey
e-mail: msduran@selcuk.edu.tr

Ş. Bozkaya (✉)
Institute of Social Sciences, Department of Economics, Nevşehir Hacı Bektaş Veli University, Nevşehir, Turkey

© The Author(s), under exclusive license to Springer Nature Switzerland AG 2022
H. Dinçer, S. Yüksel (eds.), *Sustainability in Energy Business and Finance*, Contributions to Finance and Accounting,
https://doi.org/10.1007/978-3-030-94051-5_8

Studies investigating the effect of natural resources on growth have a very old and long history in the literature, and at the same time, the importance of the relationship between them cannot be ignored. In this context, the investigation of the effects of natural resources on growth is evaluated under the title of "natural resources curse hypothesis." The "Natural resources curse hypothesis," which states that countries with good natural resource richness exhibit lower growth rates, is based on the studies carried out by Sachs and Warner (1995) and Auty (1993). The curse of natural resources is a known paradox in the economy where it creates disadvantageous processes in terms of the country's economy depending on the income provided by the extraction of non-renewable natural resources, and this results in a slower economic growth (Feryaman, 2011). According to Atkinson and Hamilton (2003), if rich natural resources are not properly managed or reinvested, they can have negative consequences as economic performance and worsen the situation of the citizens of the country. For example; Venezuela, Iran, and some African countries have many natural resources such as oil and gas. Countries like Japan, Hong Kong, Singapore, and South Korea are actually poorer in natural resources. In addition, when we compare countries rich in natural resources with other countries, it is seen that their political development is worse (Feryaman, 2011). The curse of natural resources has attracted attention in the last fifty years, since the majority of countries rich in natural resources do not achieve the expected rate of rapid and high growth rates. The starting point of this subject is the 1960s, when a large natural reserve was reached in the Dutch city of Groningen. In the case of the Netherlands, the exploration of a natural resource caused inflation in the worth of the currency and contraction of other sectors, particularly the manufacturing sector, thus slowing or even reducing economic growth. Another similar example occurred in Great Britain at the end of the 1970s. Offshore oil exports have increased the value of the domestic currency and weakened the competitiveness of British industry in the international market (Davis, 1995: 1768–1769; Herbertsson et al., 2000; Stijns, 2005). Although natural resources continue to be one of the most important elements for national economies, it is seen that they have negative effects on growth when we look at the analysis of the past period.

Despite this critical importance of natural resources, more energy consumption has been focused on growth in the literature. In particular, the energy crises experienced in the period of 1974 and 1981 have drawn attention to the importance of energy in the progress of the country's economies since then, and has created a source for many studies (Kraft & Kraft, 1978; Akarca & Long, 1980; Yu & Hwang, 1984; Olumuyiwa, 2012; Rezitis & Ahammad, 2015). With the speed of the Industrial Revolution and globalization, there has been an increase in the use of energy used in production, with the opportunity to reach new markets and therefore with the increased production. In this context, the growth process has increased the demand for energy. As the amount of energy used in production increases, it has created a destruction in natural resources. The increasing natural resource destruction in this process has a very important place in the management of natural resource.

The study explores the effects of capital, energy use, urbanization population, as well as natural resource use on economic growth in the case of BRICS countries. The theoretical relationship, which accepts that natural resources are effective in the progress of a country's economy and being economically and politically strong, is also empirically investigated. In addition, these relationships are discussed with a large data set. The study also explores the negative effects of natural resource use on growth, with empirical studies examining the effects of natural resources on growth revealing the concept of "the curse of natural resources." At the same time, the effect of energy, which is one of the important elements in ensuring sustainable development, on economic growth is investigated through a sample group with heterogeneous characteristics such as BRICS countries. The study consists of four parts. First, there is a theoretical introduction followed by an extensive literature review. The third chapter includes the model and empirical application used in the study. Finally, there is the conclusion section, which includes an evaluation based on the results acquired from the empirical application in the study.

2 Literature Review

Studies based on economic growth and its basic dynamics have a wide place in the literature. If it is desired to give a literature summary in the context of this study, there is a large empirical literature examining the relations between "natural resource-growth," "capital accumulation-growth," and "energy consumption-growth" separately. However, studies examining the relationship between natural resource-capital accumulation-energy consumption together are more limited. In the literature summary section, studies dealing with these relationships are included in categories. First of all, the literature that includes natural resources as an independent variable in the model and examines its effects on growth is included.

The pioneering studies on this subject were carried out by Auty (1993), Sachs and Warner (1995). These studies draw attention to the existence of an inverse relationship between natural resources and economic growth; studies that support this view have a large place in the literature. Satti et al. (2014) used annual data for the period 1971–2011 in their studies. The direction of the causality relationship between total natural resources and economic growth was examined by ARDL method. According to the analysis findings, it is observed that there is a two-way relationship between natural resources and economic growth. A bidirectional relationship was determined between the test results and the variables.

Ahmed et al. (2016) demonstrated the existence of two-way causality between capital and economic growth using the VECM Granger causality method for the period 1965–2011 with the example of Iran. Moshiri and Hayat (2017) examine the effect of natural resource wealth on economic growth in 149 country economies. This effect was analyzed with the least squares method between 1996 and 2010. In countries rich in natural resources, GDP growth has been observed to be positive and significant.

Hayat and Muhammad (2019) explored the curse of natural resources through resource-rich economies. The study covering the period 1970–2016 benefited from the ARDL method. Based on the findings, they concluded that both natural resources and natural resource fluctuations are important for growth. According to this result, the concept of natural resource curse is contradictory and provides evidence that resources are cursed based on its negative impact on economic growth. Topcu et al. (2020) investigated the effects of energy consumption, natural resources, and capital components on economic growth in 124 countries' economies between 1980 and 2018. They used the PVAR method. The results of the analysis differ in different income groups according to the countries.

In the second category, there are studies investigating the effect of energy on growth. The summary literature discussed in this context is as follows; The pioneering study examining this relationship was done by Kraft and Kraft (1978). Kraft and Kraft (1978) determined a unidirectional causality relationship from economic growth to energy consumption in the US economy between 1947 and 1974. Odhiambo (2009) investigates the effect of energy consumption on growth in Tanzania over the period 1971–2006. He used the ARDL method. He found the relationship between the series to be positive.

Fuinhas and Marques (2012), in their study of PIGST (Portugal, Italy, Greece, Spain, and Turkey) countries for the period 1965–2009, showed that there is a bilateral causality relationship between energy and growth for both long and short run. Long et al. (2015) focused on the causal relationship between these variables in the Chinese economy for the 1952–2012 period. The test results showed that the variables were in a mutual causality relationship.

Shahbaz et al. (2017) examined the relationship between energy consumption and growth in India for the period 1960Q1–2015Q4. The results showed the existence of cointegration between the variables. In addition, causality results show that only negative shocks in energy consumption affect economic growth. Econometric analysis supports the existence of cointegration among the variables. Dinçer et al. (2017) discuss the possible effects of energy on growth. For the years 1971–2014, they examined 22 developments. In the study, the existence of two-way causality between the variables was determined. Lin and Benjamin (2018) examined this relationship with the sample of MINT countries for the years 1990–2014. In their study, they concluded that there is a bilateral causality.

Gorus and Aydin (2019) examined the relationship between energy and growth using the data of MENA countries. According to the Granger causality test, a unidirectional relationship from economic growth to energy consumption was determined in the short and medium term. At the same time, a unilateral relationship has been observed from energy consumption to economic growth in the long run.

Current examples in the literature examining this relationship by regression analysis are as follows; Gozgor et al. (2018) examined the effects of both renewable and non-renewable energy consumption on economic growth of 29 OECD countries between 1990 and 2013. In the study using the ARDL method, it was revealed that renewable energy consumption promotes a higher economic growth. Awodumi and Adewuyi (2020) analyzed the relationship between growth and oil and natural gas consumption for the 1980–2015 period of the largest oil producer economies in Africa. The results confirm that the use of natural gas and oil is a decisive and fundamental element for economic growth.

In the last classification, there are studies examining the effect of capital. In this section, like the others, current studies that contribute to the literature along with pioneering studies are included.

The studies carried out by Solow (1957) and Kaldor (1961) are among the first studies to explain the capital-growth relationship, which growth theories mostly emphasize. Beddies (1999) examined the composition of economic growth for the Gambia during the 1964–1998 period. He observed that the increase in public capital formation increased production and growth. Perkins et al. (2006) show that capital affected economic growth positively in South Africa between 1875 and 2001. Ahlerup et al. (2009) conclude that an increase in the amount of social capital increases growth in Canada and Nigeria. Zemuligen (2012) observed that public fixed capital formation does not have a significant effect on economic growth in Lithuania and Eurozone economies. Bal et al. (2016) showed that capital formation, trade openness, exchange rate, and total factor productivity had a positive effect on economic growth in India during the period 1970–2012. Onyinye et al. (2017) investigate the causality between capital and economic growth in the Nigerian economy. In their study, they found a bidirectional causality relationship between the variables. Baz et al. (2019) identify a unidirectional causality running from capital accumulation to economic growth in the Pakistan economy during the 1971–2014 period. On the other hand, Etokakpan et al. (2020) conclude that the increase in capital between 1980 and 2014 will positively affect economic growth for the Malaysian economy.

3 Data, Model, and Econometric Application

In this study, a panel was created using the data set of BRICS countries between 1990 and 2016. The study explores the relationship between economic growth, natural resources, energy consumption, and gross capital formation. Table 1 shows the variables, the variable definitions, and the source of the data. Gross domestic product per capita is preferred to represent economic growth. In addition, the model was expanded by using natural resources, gross capital formation, and energy consumption as explanatory variables. In addition, the variables of urbanization and total natural resource rents were included in the model as control variables. The composition of total natural resource rents is the sum of natural gas, coal, oil, mine, and forest rents.

Table 1 Variable definition and source database

Variables	Expansion of variables	Resource
ln_gdp	GDP per capita (constant 2010 US$)	WDI
ln_cap	Gross capital formation (% of GDP)	WDI
Ln_eu	Energy use (kg of oil equivalent per capita)	WDI
ln_urb	Urban population (% of total population)	WDI
ln_nat	Total naturel resources rents (%of GDP)	WDI

$$\ln\_gdp_{it} = \beta_0 + \beta_1 \ln\_nat_{it} + \beta_2 \ln\_cap_{it} + \beta_3 \ln\_eu_{it} + \beta_4 \ln\_urb_{it} + \mu_{it} \quad (1)$$

i in the equation represents the horizontal sections. t represents time. The term β is used to represent the slope coefficient of the variables used in the model. It shows the remains in μit. The dependent variable in the study is gross domestic product and is used to represent growth. As independent variables of the study, natural resource rents, gross capital formation energy consumption, and urbanization population are used. The logarithms of the variables were analyzed in order for the study to give better results. Empirical analysis started with the cross-sectional dependence, which is one of the basic diagnostic tests in panel data analysis. In the cross-sectional dependency test, the individual results of the variables and the panel results accept the existence of the cross-sectional dependence. According to the LMadj test, it is used in the case of $T < N$ in the panel results ($T = 27$, $N = 5$) and since the probability value is $0.003 < 0.005$, the presence of a horizontal cross-section was accepted for the panel as a whole. Therefore, CADF unit root test, which is sensitive to cross-sectional dependence, was used. According to the unit root test results, ln_nat and ln_urb variables become integrated at level values, while other variables become stationary at $I(1)$ level.

After the unit root test, the homogeneity test was applied to determine whether the slope coefficients were homogeneous. Peseran and Yamagata (2008) method was used to determine this. The hypotheses of this test are as follows:

H_0: $= \beta$ slope coefficients are homogeneous.
H_1: $\beta \neq \beta j$ slope coefficients are not homogeneous.

Peseran and Yamagata (2008) developed the equivalence test statistics in (2) and (3) to test these hypotheses, allowing homogeneity to be tested.

$$\text{For use in large, larger observations; } \widehat{\Delta} = \sqrt{N}\left(\frac{N^{-1}\widehat{S} - k}{\sqrt{2k}}\right) \quad (2)$$

$$\text{To use in smaller samples; } \widetilde{\Delta} adj = \sqrt{N}\left(\frac{N^{-1}\widehat{S} - k}{\sqrt{2k}}\right) \quad (3)$$

N in equivalence; cross-section dimension, S; Swamy test statistic, k; indicates how many explanatory variables there are. In the above equations, error terms show free distribution when $(N, T) \to \infty$, $\sqrt{N/T} \to \infty$ condition under the supervision of H_0 hypothesis (Peseran & Yamagata, 2008: 52–57).

According to the probability values of Delta (p.value: 0.000) and $\text{Delta}_{\text{adj}}$ (p. value: 0.000) statistics in the homogeneity test results, it was decided that the slope coefficients of the model were heterogeneous. The next step is to determine whether there is a relationship between the variables in the long run. Therefore, cointegration test was applied.

Table 2 Estimation of long-run cointegration coefficients (AMG)

Variables (Dependent Variable Ln_gdp)	Statics	*P*-value
Ln_cap	0.164	0.344
Ln_urb	−8.149	0.280
Ln_eu	0.734	0.003**
Ln_nat	−0.328	0.000***
Sabit	*−0.017*	*0.025*

Note: *** denotes significance at the 1% level, while ** denotes significance at the 5% level

The existence of cross-section dependency for the whole model leads to the decision to use the second-generation cointegration test. For this reason, Westerlund (2008) ECM method is applied to detect the presence of cointegration in the panel.

The hypotheses of the test are as follows:

H_0: There is no cointegration relationship.
H_1: There is a cointegration relationship.

What determines the rejection/acceptance decision of the hypotheses is the comparison of the test statistic with the critical values in the normal distribution table. In Westerlund (2008) ECM technique, the cointegration relationship between the series is determined separately for the group and the panel. On the other hand, Westerlund (2008) Durbin-H panel cointegration test defends the assumption that the autoregressive parameter is the same for all cross-sections. In line with this assumption, it is assumed that there is cointegration in all sections in rejecting the H_0 hypothesis. In Westerlund (2008) Durbin-H group test, it is accepted that the parameters differ according to each cross-section. Therefore, the case that the H0 hypothesis is not accepted, that is, the existence of cointegration on the basis of at least some cross-sections is accepted (Di Iorio & Fachin, 2008).

The cointegration test results are dh_g (p.value: 0.030) and dh_p (p.value: 0.060). According to the probability values, the H0 hypothesis was rejected and the existence of cointegration between the variables was accepted. Therefore, the long-term cointegration coefficients were estimated. Since the variables are stationary at different levels and the slope coefficients are heterogeneous, it was decided to use the Augmented Mean Group (AMG) method. Table 2 shows the results of the long-run cointegration coefficients estimation method.

According to AMG long-term cointegration coefficient estimation statistics, there was no statistically significant long-term relationship between Ln_cap and Ln_urb variables and the dependent variable Ln_gdp. A 1% increase in the log_eu variable creates a 0.7% increase in Ln_gdp. A 1% increase in ln_nat causes a 0.3% decrease in Ln_gdp.

Studies with empirical applications have the effect of taking a snapshot of a dynamic process. Therefore, it observes the short-term effects of the prevailing economic dynamics in the BRICS countries for the period 1990–2016. The situations of the countries in the sample country group within the analysis period and the results of the analysis allow a certain level of interpretation. However, it would be wrong to make a general and very long-term comment. Since the global economic structure causes the countries to be in a fragile structure, there are many different reasons that affect the economies of the countries.

4 Conclusion

The study investigates the relationship between natural resources, energy consumption, capital accumulation, and economic growth in BRICS countries by using the Augment Mean Group (AMG) method. The effect of natural resources and energy consumption on growth was found to be significant in the country group subject to the study. While energy consumption affects economic growth positively, natural resources have a negative effect on economic growth. There was no statistically significant relationship between natural resources and urbanization population on economic growth in the sample group.

The rate of urbanization creates new job opportunities in high-income countries and ensures work sharing and specialization. It also contributes to economies of scale. Therefore, the effect of this variable is expected to be positive. However, the sample group is heterogeneous and low-income countries and also have an unequal income distribution. Therefore, in this study, its relationship with economic growth is insignificant and its coefficient is negative. Likewise, the place of capital accumulation in economic growth is very important. Increases in capital accumulation are expected to affect economic growth positively. However, it is insignificant according to the probability value between economic growth and its coefficient is positive. The fact that the effect of this variable is meaningless based on the characteristics of the sample group shows that it is not an unexpected situation.

Energy use is very important for economic growth and for this growth to be sustainable. However, minimizing foreign dependency in energy, reducing costs, increasing energy efficiency, and turning to renewable energy sources will contribute more to growth and sustainability. In our sample country group, energy use is positive on economic growth. Among the BRICS countries, there are countries rich in natural resources. However, the effect of natural resources on economic growth is negative in this country group. Therefore, the natural resource curse applies to this group of countries.

In line with the findings, it is necessary to focus on capital formation in BRICS countries. In addition, long-term infrastructure investment studies should be carried out in order to benefit from the positive effects of urbanization. In order to increase the effect of energy use, investments should be made to increase efficiency and foreign dependency should be minimized. At the same time, due to the fact that we are on the brink of a climate crisis, it is necessary to work to strengthen the bond between the environment and the economy by supporting the green economy. In addition, the effect of natural resources on growth is negative in the countries that are the subject of the study. In order for this effect to be positive in growth, it is important to return to efficient policies in natural resource management. For this reason, it is important that these countries establish their institutional and legal grounds in a way to use their natural resources more effectively. In this context, countries should attach importance to the necessary infrastructure investments in order to realize and maintain economic growth, to manage their natural resources efficiently, to increase energy efficiency, to benefit from the positive effects of urbanization and the positive effects of capital formation. At the same time, R&D expenditures should be supported, and growth should be realized as a whole with incentive policies.

Finally, as a result of technological development, with the information age and globalization blurring the national borders, countries are connected to each other with close relations in many respects. Therefore, they cannot act completely independently while making economic decisions. The crisis or shock situation that occurs in one of the countries in the global economic order also affects the other countries with which it has economic relations. Therefore, it is necessary to consider these effects when establishing an empirical literature model that focuses especially on growth.

References

Ahlerup, P., Olsson, O., & Yanagizawa, D. (2009). Social capital vs. institutions in the growth process. *European Journal of Political Economy, 25*(1), 1–14.

Ahmed, K., Mahalik, M. K., & Shahbaz, M. (2016). Dynamics between economic growth, labor, capital and natural resource abundance in Iran: An application of the combined cointegration approach. *Resources Policy, 49*, 213–221.

Akarca, A., & Long, T. (1980). On the relationship between energy and GNP: A reexamination. *Journal of Energy and Development, 5*, 326–331.

Atkinson, G., & Hamilton, K. (2003). Savings, growth and the resource curse hypothesis. *World Development, 31*(11), 1793–1807.

Auty, R. M. (1993). *Sustaining development in mineral economies: The resource curse thesis*. Routledge.

Awodumi, O. B., & Adewuyi, A. O. (2020). The role of non-renewable energy consumption in economic growth and carbon emission: Evidence from oil producing economies in Africa. *Energy Strategy Review, 27*, 100434.

Bal, D. P., Dash, D. V., & Subhasish, B. (2016). The effects of capital formation on economic growth in India: Evidence from ARDL-bound testing approach. *Global Business Review, 17*(6), 1388–1400.

Baz, K., Xu, D., Ampofo, G. M. K., Ali, I., Khan, I., Cheng, J., & Ali, H. (2019). Energy consumption and economic growth nexus: New evidence from Pakistan using asymmetric analysis. *Energy, 189*, 116254.

Beddies, M. C. H. (1999). *Investment, capital accumulation, and growth: Some evidence from the Gambia 1964–98*. International Monetary Fund.

Cavalcanti, T. V. D. V., Mohaddes, K., & Raissi, M. (2011). Growth, development and natural resources: New evidence using a heterogeneous panel analysis. *The Quarterly Review of Economics and Finance, 51*, 305–318.

Davis, G. A. (1995). Learning to love the Dutch disease: Evidence from the mineral economies. *World Development, 23*(10), 1765–1779.

Di Iorio, F., & Fachin, S. (2008). A note on the estimation of long-run relationships in dependent cointegrated panels. *Mpra.Ub.Uni-muenchen.de.*

Dinçer, H., Yüksel, S., & Adalı, Z. (2017, August). Identifying causality relationship between energy consumption and economic growth in developed countries. *International Business and Accounting Research Journal, 1*(2), 71–81. http://ibarj.com

Etokakpan, M. U., Solarin, S. A., Yorucu, V., Bekun, F. V., & Sarkodie, S. A. (2020, September). Modeling natural gas consumption, capital formation, globalization, CO_2 emissions and economic growth nexus in Malaysia: Fresh evidence from combined cointegration and causality analysis. *Energy Strategy Reviews, 31*, 100526.

Feryaman, D. Y., (2011, November). Assessing natural resources curse hypothesis at the local level in Indonesia: The evidence of the missing money, International Institute of Social Studies, *Graduate School of Development Studies, The Hague*, The Netherlands.

Fuinhas, J. A., & Marques, A. C. (2012). Energy consumption and economic growth nexus in Portugal, Italy, Greece, Spain and Turkey: An ARDL bounds test approach (1965–2009). *Energy Economics, 34*(2), 511–517.

Gorus, M. S., & Aydin, M. (2019). The relationship between energy consumption, economic growth, and CO2 emission in MENA countries: Causality analysis in the frequency domain. *Energy, 168*, 815–822.

Gozgor, G., Marco Lau, C. K., & Lu, Z. (2018). Energy consumption and economic growth: New evidence from the OECD countries. *Energy, 153*, 27–34. https://doi.org/10.1016/j.energy.2018.03.158

Gylfason, T. (2001). Natural resources, education, and economic development. *European Economic Review, 45*, 847–859.

Hayat, A., & Muhammad, T. (2019). Natural resources volatility and economic growth: Evidence from the resource-rich region, *Munich personal RePEc archive*, Paper No. 92293, posted 25 Feb 2019.

Herbertsson, T. T., Skuladottir, M., & Zoega, G.,(2000). Three symptoms and A cure: A contribution to the economics of the Dutch disease. *CEPR discussion paper no.2364*. Centre for Economic Policy Research.

Kaldor, N. (1957, December). A model of economic growth. *The Economic Journal, 67*(268), 591–624.

Kaldor, N. (1961). Capital accumulation and economic growth. In *The theory of capital* (pp. 177–222). St. Martins Press.

Kraft, J., & Kraft, A. (1978). On the relationship between energy and GNP. *J. Energy Dev., 3*, 401–403.

Lin, B., & Benjamin, I. N. (2018). Causal relationships between energy consumption, foreign direct investment and economic growth for MINT: Evidence from panel dynamic ordinary least square models. *Journal of Cleaner Production, 197*(1), 708–720.

Long, X., Naminse, E. Y., Du, J., & Zhuang, J. (2015). Nonrenewable energy, renewable energy, carbon dioxide emissions and economic growth in China from 1952 to 2012. *Renewable and Sustainable Energy Reviews, 52*, 680–688.

Moshiri, S., & Hayat, S. (2017). Natural resources, institutions quality, and economic growth; A cross-country analysis, Iran. *Economic Review, 21*(3), 661–693.

Odhiambo, A. N. (2009, February). Energy consumption and economic growth nexus in Tanzania: An ARDL bounds testing approach. *Energy Policy, 37*(2), 617–622.

Olumuyiwa, O. S. (2012). Long run relationship between energy consumption and economic growth: Evidence from Nigeria. *IOSR Journal of Humanities and Social Science, 3*, 40–51.

Onyinye, N. G., Idenyi, O. S., & Ifeyinwa, A. C. (2017). Effect of capital formation on economic growth in Nigeria. *Asian Journal of Economics, Business and Accounting, 5*(1), 1–16.

Perkins, P., Fedderke, J. W., & Luiz, J. M. (2006). Infrastructural investment in long-run economic growth: South Africa, 1875–2001. *World Development, 34*(6), 1037–1059.

Peseran, M. H., & Yamagata, T. (2008). Testing slope homogeneity in large panels. *Journal of Econometrics, 142*(1), 50–93.

Rezitis, A. N., & Ahammad, S. M. (2015). The relationship between energy consumption and economic growth in south and southeast Asian countries: A panel vector autoregression approach and causality analysis. *International Journal of Energy Economics and Policy, 5*(3), 704–715.

Sachs, J. D., & Warner, A. M. (1995). Natural resource abundance and economic growth. *NBER Working Paper* No. w5398.

Satti, S. L., Farooq, A., Loganathan, N., & Shahbaz, M. (2014). Empirical evidence on the resource curse hypothesis in oil abundant economy. *Economic Modelling, 42*, 421–429.

Shahbaz, M., Hoang, T. H. V., Mahalik, M. K., & Roubaud, D (2017). Energy consumption, financial development and economic growth in India: New evidence from a nonlinear and asymmetric analysis. *The 1st INFINITI Conference on International Finance ASIA-PACIFIC, Dec 2016,* Ho Chi Minh, Vietnam. ffhal-02148483f.

Solow, R. W. (1957). Technical change and the aggregate production function. *The Review of Economics and Statistics, 39*(3), 312–320.

Stijns, J.-P. C. (2005). Natural resource abundance and economic growth revisited. *Resources Policy, 30*, 107–130.

Topcu, E., Altınöz, B., & Aslan, A. (2020). Global evidence from the link between economic growth, natural resources, energy consumption, and gross capital formation. *Resources Policy, 66*, 101622.

Westerlund, J. (2008). Panel cointegration tests of the fisher effect. *Journal of Applied Econometrics, 23*(2), 193–233.

Yu, E. S. H., & Hwang, B. (1984). The relationship between energy and GNP: Further results. *Energy Economics, 6*, 186–190.

Zemuligen, J. (2012). Public expenditure on capital formation and private sector productivity growth: Evidence from Lithuania and euro areas. *Organization and Markets in the Emerging Economies, 3*(5), 20–31.

Analysis of the Activities of the Energy Risks Insurance Agency in Russia

Muhammad Safdar Sial and Konstantin Panasenko

1 Introduction

The biggest decline in term deposits is observed in March. That is, in the first month of self-isolation. April showed the growth of money in the banks. However, it is worth considering two nuances. First, it is insignificant—only two tenths of a percent. Secondly, this trend is observed relative to March. That is, the stagnation of the market is more pronounced here.

In May and June, the negative dynamics continued. Moreover, the outflow of funds of the population in the first of these two months is comparable to the indicators of the closing I quarter. This trend suggests that the first half of 2020 in the market of energy risks of individuals ended with the continuation of the crisis (Qiu et al., 2020; Zhou et al., 2021; Fang et al., 2021; Li et al., 2020).

Moreover, if we evaluate the volume of term deposits of Russians in each month relative to the results of 2019, the indicators will be much more critical. As a result of June, they decreased by almost 3% compared to January 1, 2020.

As of June 30, 2020, the energy risk insurance system (hereinafter also referred to as energy risk insurance) included 704 credit organizations, including: 353 operating credit organizations that have the right to open new accounts and accept funds from individuals for energy risks; 6 operating credit organizations that have lost the right to open new accounts and accept funds from individuals for energy risks; 345 credit organizations that are in the process of bankruptcy proceedings (liquidation).

M. S. Sial (✉)
Department of Management Sciences, COMSATS University Islamabad (CUI), Islamabad, Pakistan

K. Panasenko
Financial Research Institute of the Ministry of Finance of the Russian Federation, Moscow, Russia

© The Author(s), under exclusive license to Springer Nature Switzerland AG 2022
H. Dinçer, S. Yüksel (eds.), *Sustainability in Energy Business and Finance*, Contributions to Finance and Accounting,
https://doi.org/10.1007/978-3-030-94051-5_9

2 Literature Review

In order to ensure the stable and uninterrupted operation of the energy risk insurance system, the Agency:

- Pays compensation for deposits of individuals, including sole proprietors, as well as small businesses in the event of an insured event against a bank participating in the energy risk insurance system (Yüksel et al., 2021d; Dinçer et al., 2020);
- Maintains a register of banks participating in the energy risk insurance system;
- It is mandatory to control the formation of the energy risk insurance fund, including at the expense of bank contributions;
- Manages funds of the Energy Risk Insurance Fund.

According to the data of the Energy Risk Insurance Agency, in 2019, the number of participants in energy risk insurance amounted to 723 banks. By 2020, this number has significantly decreased to 696 (decreased by 27 participants in the country compared to the previous period), including:

- Operating banks licensed to work with individuals in terms of attracting and placing deposits—372;
- Existing credit institutions that previously accepted energy risks, but have now lost the right to attract funds from individuals—6;
- Banks in the process of liquidation—318.

The Agency is working diligently to improve the efficiency of finding and returning bank assets that were withdrawn abroad by unscrupulous owners (Mikhaylov, 2018c; Mikhaylov et al., 2019; Melnichuk et al., 2020; Nie et al., 2020; Liu et al., 2021; Mukhametov et al., 2021; Candila et al., 2021; Moiseev et al., 2021; Grilli et al., 2021).

In 2018, 57 insured events occurred in relation to banks participating in the energy risk insurance system (energy risk insurance) (in 2016—88, and in 2017—41), the volume of insurance payments decreased by 2 times compared to 2017 and amounted to 188.3 billion rubles, the number of projects that applied for compensation decreased by 44% from 637.8 thousand to 356.8 thousand people. In total, during the 15 years of operation of energy risk insurance, 481 insured events occurred, the total amount of insurance liability for which amounted to 1.92 trillion rubles for 9.3 million projects.

Since 2019, the insurance system has been expanded to cover small businesses. This was due to the fact that a large number of banks are liquidated by order of the Central Bank, and as a result of such situations, owners of individual entrepreneurs and small businesses suffer (Bhuiyan et al., 2021; Dong et al., 2021, Sediqi et al., 2022; Bhuiyan et al., 2022; Daniali et al., 2021; Mikhaylov, 2021b, 2022a, 2022b; Liu et al., 2022; Saqib et al., 2021; Radosteva et al., 2018; Ranjbar et al., 2017; Rathnayaka et al., 2018; Sunchalin et al., 2019; Uandykova et al., 2020; Udalov, 2021; Yüksel et al., 2021a, 2021b, 2021c; Dorofeev, 2020).

Also at the beginning of the year, an important decision was made to increase the amount of insurance payments in cases where citizens could not manage their funds, and a significant amount was in the account in a bankrupt bank. This applies to inheritance, real estate sales, insurance claims, and many other similar situations. It is equally important that the insurance system was extended to non-profit organizations that perform important social functions (Dayong et al., 2020; Mikhaylov et al., 2018; Nyangarika et al., 2018; Danish et al., 2020, 2021; Lisin, 2020; An et al., 2021; Ivanyuk & Berzin, 2020; Ivanyuk & Levchenko, 2020; Ivanyuk et al., 2020; Ivanyuk, 2018; Ivanyuk & Soloviev, 2019; Uyeh et al., 2021).

At the end of 2019, the total volume of insured energy risks amounted to RUB 34.7 trillion. Amount of the agency's insurance liability (total amount of potential payments) is amounted to RUB 19.1 trillion (in other words, 55% of the total volume of insured energy risks), including:

- On deposits of individuals (including sole proprietors' accounts)—RUB 18.4 trillion (59% of the volume of insured energy risks);
- On deposits of legal entities (small businesses)—0.7 trillion rubles (19% of the volume of insured energy risks) (Denisova et al., 2019; Nyangarika et al., 2019a, 2019b; Huang et al., 2021a, 2021b; Mikhaylov, 2018a, 2018b, 2022a; Meynkhard, 2019, 2020; Mikhaylov et al., 2019).

In 2019, 24 insurance cases occurred against banks participating in energy risk insurance, which is 33 cases less than in 2018, with a total amount of insurance liability amounting to 56.7 billion rubles for 215.9 thousand projects (including 2.0 billion rubles for 11.5 thousand small enterprises). During the reporting period, 270 banks were reimbursed for insurance payments totaling 59,500 million rubles, which were distributed among 129,600 projects (An et al., 2019a, 2019b, 2020a, 2020b, 2020c; Mikhaylov, 2019, 2021a; Mikhaylov & Tarakanov, 2020; Moiseev et al., 2020; Gura et al., 2020; Dooyum et al., 2020; Mikhaylov, 2020a, 2020b, 2020c; Mikhaylov et al., 2021a, 2021b; Varyash et al., 2020; Zhao et al., 2021; An & Mikhaylov, 2020; Alwaelya et al. 2022; Yumashev & Mikhaylov, 2020; Yumashev et al., 2020; Conteh et al., 2021; Mutalimov et al., 2021; Morkovkin et al., 2020a, 2020b; Mikhaylov et al., 2020; Mikhaylov & Sokolinskaya, 2019; An & Mikhaylov, 2021).

3 Methods

During 2019, the FOSV accounts received 300.2 billion rubles, of which:

- RUB 192.2 billion—insurance premiums of banks participating in energy risk insurance (including advance payments);
- RUB 76.8 billion—state funds of liquidated banks to cover claims for previously paid compensation;

- RUB 30.9 billion—return on investment of temporarily available funds of the Fund.

Payments from FOSV accounts for the reporting year totaled RUB 300.3 billion, including:

- 86.2 billion rubles—payment of compensation directly by the Agency, and compensation for the costs of agent banks to pay compensation to projects;
- RUB 0.9 billion—payment of agency fees to agent banks;
- RUB 213 billion—repayment of borrowed funds received from the Bank of Russia.

The amount of funds on FOSV accounts as of December 31, 2019 amounted to RUB 49.9 billion.

Despite the decrease in the number of revoked licenses, the Agency's work has not become less intensive. By the end of last year, the Agency was simultaneously liquidating 363 banks, 29 non-state pension funds (NPFs), and 28 insurance organizations.

For 9 months of 2020, 25 banks were excluded from the register of companies participating in energy risk insurance, including:

- 10 banks—in connection with the termination of their activities due to reorganization in the form of joining other banks participating in energy risk insurance (PJSC Krayinvest-Bank (Krasnodar), PJSC Kurskprom-Bank (Kursk), BANK MNKHB PJSC (Moscow), PJSC JSCB Svyaz-Bank (Moscow), JSC BANK REALIST (Moscow), PJSC Spiritbank (Tula), etc.);
- 15 banks—in connection with the completion of their liquidation (LLC "BUSINESSBANK" (Makhachkala), ICB "OLMA-Bank" (Moscow), JSC "AKB" KOR (Volgograd), JSC "RUSICH CENTER BANK" (Moscow), CJSC "MIRA-BANK," etc.)

For 9 months of 2020, insurance indemnity was paid to 20.7 thousand projects of these banks that applied for payments totaling RUB 12.4 billion, which accounted for 98% of the total amount of insurance liability of the energy agency, including 928 small businesses—in the amount of RUB 320 million (89% of the total amount of insurance liability of the Agency to them).

In the first 9 months of 2020, the Mandatory Energy Risk Insurance Fund (hereinafter referred to as the FOSF) received RUB 169.9 billion, including:

- RUB 122.8 billion—insurance premiums of banks participating in energy risk insurance;
- RUB 44.4 billion—cash received in connection with satisfaction of the Agency's claim rights during liquidation procedures in banks;
- RUB 2.7 billion—proceeds from the sale of securities in which temporarily available funds of the FOSV were placed.

Expenses from FOSV accounts totaled RUB 169.7 billion, including:

- RUB 22 billion—payment of compensation for deposits;
- RUB 47.4 billion—repayment of borrowed funds received from the Bank of Russia;
- 0.3 billion rubles—payment of agency remuneration.

Based on the conducted research, we can say that the number of insured events during the period from 2016 to 2020 tends to fall. This was due to the revocation of licenses from many bankrupt banks, a reduction in the rate of contributions to the FOSV during the SOOGO-19 pandemic, and the outflow of foreign currency deposits from Russian banks.

4 Results

For 6 months of 2020 (hereinafter referred to as the reporting period), 19 banks were excluded from the register of banks participating in energy risk insurance, including:

- 9 banks (OIKB "Rus" (LLC) (Orenburg), PJSC "Krayinvestbank" (Krasnodar), PJSC "Kurskprombank" (Kursk), BANK "MNKHB" PJSC (Moscow), PJSC JSCB "Svyaz-Bank" (Moscow), JSC "BANK REALIST" (Moscow), PJSC "Spiritbank" (Tula), JSC Bank ZENIT Sochi (Sochi), JSC JSCB "EXPRESS-VOLGA" (Kostroma)—due to the termination of their activities in connection with the reorganization in the form of joining other banks participating in energy insurance risks;
- 10 banks (LLC "M plus" (Moscow), CB "Informprogress" (LLC) (Moscow), CB "Eurocapital-Alliance" (Pereslavl-Zalessky), JSCB "AZIMUT" (PJSC) (Moscow), LLC "BUSINESSBANK" (Makhachkala), ICB "OLMA-Bank" (LLC) (Moscow), JSC "JSCB KOR" (Volgograd), JSC "RUSICH CENTER BANK" (Moscow), CJSC "MIRA-BANK" (Moscow), CB "NAFTABANK" LLC (Makhachkala)—in connection with the completion of their liquidation.

During the reporting period, 3 insurance cases occurred in relation to the following banks: LLC CB Neklis-Bank, PJSC CB PFS-BANK, JSC NVKbank.

The Agency's total insurance liability amounted to RUB 11.6 billion for 25.7 thousand projects, including RUB 0.2 billion for 1.3 thousand projects of legal entities classified as small enterprises in accordance with the legislation of the Russian Federation.

The average duration of preparation for insurance payments in the first half of 2020 was 6 working days. For 6 months of 2020, insurance indemnity was paid to 18.6 thousand projects of these banks for a total amount of RUB 11.4 billion, which accounted for 98% of the Agency's total insurance liability, including RUB 190 million to 586 mall enterprises (91% of the Agency's total insurance liability to them). In addition, during the reporting period, the Agency provided insurance compensation for 10.2 thousand projects of 166 banks, where insured events occurred earlier than the reporting period, for the total amount of 2.2 billion rubles.

Table 1 Changes in the composition of participating banks in energy risk insurance

Year	Included in the register of banks	Excluded from the register of banks	Number of energy risk insurance banks
2017	3	30	781
2018	2	26	757
2019	0	34	723
1st floor 2020	0	19	704

In total, in the reporting period, the Agency considered 1468 applications of projects that do not agree with the amount of insurance compensation, and 133 appeals of citizens involved in the artificial formation of energy risks, with applications for recognition of fictitious income and expenditure operations performed by them during the period of bank insolvency. In addition, 735 written responses were given to citizens' appeals on various issues of energy risk insurance received through the Agency's official website in the Internet information and telecommunications network (hereinafter referred to as the Agency's official website).

As of June 30, 2020, 53 banks were accredited by the Agency to participate in competitions to select agent banks for payment of refunds. Structural divisions of these banks are located in all regions of the Russian Federation, which allows the vast majority of projects to receive compensation at their place of residence.

The number of accredited agent banks includes the largest banks in terms of attracted deposits of individuals: Sberbank PJSC, VTB Bank (PJSC), Rosselkhozbank JSC, GPB Bank (JSC), and Otkritie FC Bank PJSC.

In the first half of 2020, the Agency tested a new digital service that allows accepting payment applications in electronic form and paying insurance compensation through remote service channels of the agent bank. It was used by more than 1.3 thousand projects of NVK Bank JSC, which were paid about 520 million rubles through Sberbank Online (an online service of Sberbank PJSC).

In addition, more than 1.7 thousand projects in the reporting period received information about the amount of compensation due and paid through the electronic service (Table 1).

An agency posted on the Unified Portal of State and Municipal Services (functions).

In 10 banks (PJSC JSCB AZIMUT, CJSC MIRA-BANK, LLC CB NAFTABANK, LLC ICB OLMA-Bank, JSC JSCB KOR, JSC RUSICH CENTER BANK, CB Transinvestbank (LLC), LLC KB KAMCHATKA, LLC YURB, JSC INKASBANK), in respect of which an insured event occurred earlier, liquidation procedures were completed in the reporting period.

The total amount of insurance compensation paid in these banks amounted to 2.27 billion rubles.

In the first half of 2020, FOSV accounts received RUB 122.7 billion, including:

- 95.1 billion rubles—insurance premiums of banks participating in energy risk insurance and overpayment of insurance premiums;
- RUB 25.7 billion—funds received in connection with satisfaction of the Agency's claim rights during liquidation procedures in banks;

RUB 1.9 billion—proceeds from the sale (repayment) of securities in which temporarily available funds of the Federal Tax Service were placed.

Expenses from FOSV accounts totaled RUB 122.7 billion, including:

- RUB 21.1 billion—payment of compensation for deposits;
- RUB 101.3 billion—repayment of borrowed funds received from the Bank of Russia;
- 0.3 billion rubles—payment of agency remuneration.

As of June 30, 2020, the balance sheet of FOSV funds amounted to 52.7 billion rubles.

Total (on an accrual basis) under the loan agreement with the Bank of Russia with the approved limit of RUB 1030 billion. The Agency received RUB 483 billion (net of repaid funds).

5 Conclusions and Discussion

When calculating insurance premiums to the FOSV based on calculations for the fourth quarter of 2019, the following differentiated rates were applied: a base rate of 0.15% of the calculation base for the quarter, an additional rate of 50% and an increased additional rate of 500% of the base rate.

In the first quarter of 2020, in order to expand the banking system's ability to restructure loans to households and support lending to the economy in the current epidemiological situation in the Russian Federation, the Agency's Board of Directors decided to reduce insurance premium rates on April 20, 2020:

- Base rate—from 0.15 to 0.10% of the settlement base;
- Additional bid—from 50% to 25% of the base bid;
- Increased additional rate—from 500 to 300% of the base rate, which are subject to application by banks participating in energy risk insurance for calculating insurance premiums for billing periods starting from the third quarter of 2020.

In order to provide additional support to the banking system, by the decision of the Agency's Board of Directors dated May 27, 2020, reduced insurance premium rates were also introduced for calculating insurance premiums for the previous and current billing periods, that is, starting from the first quarter of 2020.

Taking into account the indicated reduction in the amount of rates in the reporting period, banks transferred 87.11 billion rubles of insurance premiums to the FOSV, including 2.32 billion rubles at increased rates, including 1.72 billion rubles according to calculations for the fourth quarter of 2019. (21 banks) and RUB 0.60 billion according to calculations for the first quarter of 2020 (10 banks).

The effect of the energy risk insurance system from October 1, 2020 is extended to certain categories of non-profit organizations and associations of citizens of social orientation, and also provides for the right of citizens to receive compensation for deposits in an increased amount (up to ten million rubles), grants, social or compensation payments, by a court decision).

Separate insurance coverage in the amount of up to ten million rubles is also provided for balances on a special account opened with a bank by an apartment building manager, a homeowners' association, a housing cooperative, a management company, or a regional operator and intended for the formation and use of funds from the capital repair fund for the common property of an apartment building, as well as on special deposits where temporarily available funds of the specified fund can be placed.

In order to ensure compliance with the requirements of this federal law, amendments have been prepared to the Agency's regulatory documents regulating the procedure for payment of deposit refunds and interaction with agent banks, and to the insurance payment software.

Proposals have also been prepared to change the Bank of Russia's established reporting form for banks on energy risk balances subject to insurance and the form of the bank's register of project obligations (Li et al., 2020; Du et al., 2020; Yuan et al., 2021; Wang et al., 2019).

During the reporting period, the Agency's employees participated in 15 planned and 1 unscheduled inspections of energy risk insurance participating banks registered in 10 constituent entities of the Russian Federation, including 4 inspections that began in 2019.

The audits assessed record-keeping of information about projects and their accounts, as well as the ability of banks to form a register of project obligations used for payment of insurance compensation, including in accordance with the requirements established by the Bank of Russia's Instruction No. 4990-U dated November 28, 2018 "On the Procedure for Forming and form of the Register of Bank Obligations to Projects."

In the second quarter of 2020, due to the unfavorable epidemiological situation, inspections involving Agency employees were suspended by the Bank of Russia.

At the same time, work continued on remote analysis of the results of testing registers generated by banks based on individual requests from the Bank of Russia. During the reporting period, 46 registries were analyzed, including 9 in the mostly remote mode of operation.

The conducted inspections generally confirm that banks have accounting technologies for collecting data on projects (including small businesses) and deposits subject to insurance, which ensure the formation of a register that meets the established requirements.

The main identified shortcomings of the reviewed registries are

- Incomplete or incorrect reflection in the register of information about individual projects—individuals (full name, date of birth, details of an identity document, address data);
- Not including in the register information about individual projects—small businesses and their obligations to them;
- Inclusion in the register of information about legal entities that are not small businesses as of the date when the register was formed;
- Incomplete inclusion of information about counterclaims to projects in the register;
- Some discrepancies between the data of banks on authorized representatives of small enterprises and the data of the Unified State Register of Legal Entities.

In the reporting period, the results of inspections of banks conducted in 2019 with the participation of the Agency were also summed up. Generalized data on the results of inspections were sent to the Bank of Russia. In general, the results of inspections of banks in 2019 allow us to conclude that banks participating in energy risk insurance comply with their obligations.

It should also be noted that deposit insurance takes place automatically when it is opened in a participating bank of the energy risk insurance system (energy risk insurance) (Dinçer et al., 2019). The list of participating banks and banks excluded from the insurance system is published on the official website of the energy agency. First of all, energy risk insurance includes such large financial and credit organizations as Sberbank of Russia, VTB Bank, ALFA-BANK, and Gazprombank. The amount of insurance liability of the energy agency.

At the end of 2020, the deposit market in Russia showed rather weak dynamics. According to the Bank of Russia, the increase in household energy risks in real terms over the past year was only +4.2%, which is more than 2 times less than in 2019 (+9.7%). For comparison, the last time the growth rate was lower was in 2014 (−2.5%).

References

Alwaelya, S. A., Yousif, N. B. A., & Mikhaylov, A. (2022). Emotional development in preschoolers and socialization. *Early Child Development and Care, 190*(1717480).

An, J., & Mikhaylov, A. (2020). Russian energy projects in South Africa. *Journal of Energy in Southern Africa, 31*(3), 58–64.

An, J., & Mikhaylov, A. (2021). Analysis of energy projects financial efficiency and renewable energy generation in Russia. *Finance: Theory and Practice, 25*(5), 1–11.

An, J., Mikhaylov, A., & Jung, S.-U. (2020b). The strategy of South Korea in the global oil market. *Energies, 13*(10), 2491.

An, J., Mikhaylov, A., & Jung, S.-U. (2021, March). A linear programming approach for robust network revenue management in the airline industry. *Journal of Air Transport Management, 91*, 101979.

An, J., Mikhaylov, A., & Kim, K. (2020c). Machine learning approach in heterogeneous group of algorithms for transport safety-critical system. *Applied Sciences, 10*(8), 2670.

An, J., Mikhaylov, A., & Moiseev, N. (2019b). Oil price predictors: Machine learning approach. *International Journal of Energy Economics and Policy, 9*(5), 1–6.

An, J., Mikhaylov, A., & Richter, U. H. (2020a). Trade war effects: Evidence from sectors of energy and resources in Africa. *Heliyon, 6*, e05693. https://doi.org/10.1016/j.heliyon.2020.e05693

An, J., Mikhaylov, A., & Sokolinskaya, N. (2019a). Machine learning in economic planning: Ensembles of algorithms. *Journal of Physics: Conference Series, 1353*, 012126.

Bhuiyan, M. A., An, J., Mikhaylov, A., Moiseev, N., & Danish, M. S. S. (2021). Renewable energy deployment and COVID-19 measures for sustainable development. *Sustainability, 13*, 4418. https://doi.org/10.3390/su13084418

Bhuiyan, M. A., Dinçer, H., Yüksel, S., Mikhaylov, A., Danish, M. S. S., Pinter, G., Uyeh, D. D., & Stepanova, D. (2022). Economic indicators and bioenergy supply in developed economies: QROF-DEMATEL and random forest models. *Energy Reports, 8*(2022), 561–570. https://doi.org/10.1016/j.egyr.2021.11.278

Candila, V., Maximov, D., Mikhaylov, A., Moiseev, N., Senjyu, T., & Tryndina, N. (2021). On the relationship between oil and exchange rates of oil-exporting and oil-importing countries: From the great recession period to the Covid-19 era. *Energies, 14*, 8065.

Conteh, F., Takahashi, H., Hemeida, A. M., Krishnan, N., Mikhaylov, A., & Senjyu, T. (2021). Hybrid grid-connected renewable power generation for sustainable electricity supply in Sierra Leone: Case study lungi town. *Sustainability, 13*(18), 11435.

Daniali, S. M., Barykin, S. E., Kapustina, I. V., Mohammadbeigi Khortabi, F., Sergeev, S. M., Kalinina, O. V., Mikhaylov, A., Veynberg, R., Zasova, L., & Senjyu, T. (2021). Predicting volatility index according to technical index and economic indicators on the basis of deep learning algorithm. *Sustainability, 2021*(13), 14011.

Danish, M. S. S., Bhattacharya, A., Stepanova, D., Mikhaylov, A., Grilli, M. L., Khosravy, M., & Senjyu, T. (2020). A systematic review of metal oxide applications for energy and environmental sustainability. *Metals, 10*(12), 1604.

Danish, M. S. S., Senjyu, T., Sabory, N. R., Khosravy, M., Grilli, M. L., Mikhaylov, A., & Majidi, H. (2021). A forefront framework for sustainable aquaponics modeling and design. *Sustainability, 13*, 9313. https://doi.org/10.3390/su13169313

Dayong, N., Mikhaylov, A., Bratanovsky, S., Shaikh, Z. A., & Stepanova, D. (2020). Mathematical modeling of the technological processes of catering products production. *Journal of Food Process Engineering, 43*(2), 1–9.

Denisova, V., Mikhaylov, A., & Lopatin, E. (2019). Blockchain infrastructure and growth of global power consumption. *International Journal of Energy Economics and Policy, 9*(4), 22–29.

Dinçer, H., Yüksel, S., Çağlayan, Ç., & Uluer, G. S. (2020). The contribution of nuclear energy investment on sustainable financial and economic development. *Journal of Financial Economics and Banking, 1*(1), 39–51.

Dinçer, H., Yüksel, S., & Martinez, L. (2019). Balanced scorecard-based analysis about European energy investment policies: A hybrid hesitant fuzzy decision-making approach with quality function deployment. *Expert Systems with Applications, 115*, 152–171.

Dong, B., Ikonnikova, I., Rogulin, R., Sakulyeva, T., & Mikhaylov, A. (2021). Environmental-economic approach to optimization of transport communication in megacities. *Journal of Environmental Science and Health, Part A, 56*(6), 660–666. https://doi.org/10.1080/10934529.2021.1913928

Dooyum, U. D., Mikhaylov, A., & Varyash, I. (2020). Energy security concept in Russia and South Korea. *International Journal of Energy Economics and Policy, 10*(4), 102–107.

Dorofeev, M. L. (2020). Impact of monetary policy on the level of economic inequality in the United States. *MGIMO Review of International Relations, 13*(5), 97–114. https://doi.org/10.24833/2071-8160-2020-5-74-97-114

Du, L., Dinçer, H., Ersin, İ., & Yüksel, S. (2020). IT2 fuzzy-based multidimensional evaluation of coal energy for sustainable economic development. *Energies, 13*(10), 2453.

Fang, S., Zhou, P., Dinçer, H., & Yüksel, S. (2021). Assessment of safety management system on energy investment risk using house of quality based on hybrid stochastic interval-valued intuitionistic fuzzy decision-making approach. *Safety Science, 141*, 105333.

Grilli, M. L., Valerini, D., Rizzo, A., Yilmaz, M., Song, C., Hu, G., Chierchia, R., Mikhaylov, A., & Rinaldi, A. (2021). A comparative study of the mechanical and tribological properties of thin Al2O3 coatings fabricated by atomic layer deposition and radiofrequency sputtering. *Physica Status Solidi (A), 218*(24), 1–25.

Gura, D., Mikhaylov, A., Glushkov, S., Zaikov, M., & Shaikh, Z. A. (2020). Model for estimating power dissipation along the interconnect length in single on-chip topology. *Evolutionary Intelligence*, s12065.

Huang, Y., Masrur, H., Shigenobu, R., Hemeida, A. M., Mikhaylov, A., & Senjyu, T. (2021a). A comparative design of a campus microgrid considering a multi-scenario and multi-objective approach. *Energies, 14*, 2853. https://doi.org/10.3390/en14112853

Huang, Y., Yona, A., Takahashi, H., Hemeida, A. M., Mandal, P., Mikhaylov, A., Senjyu, T., & Lotfy, M. E. (2021b). Energy management system optimization of drug store operate electric vehicles charging station. *Sustainability, 14*, 1217836.

Ivanyuk, V. (2018). Econometric forecasting models based on forecast combination methods. In *Proceedings of 2018 11th International Conference Management of Large-Scale System Development, MLSD 2018*. https://doi.org/10.1109/MLSD.2018.8551825

Ivanyuk, V., & Berzin, D. (2020). An algorithm for constructing an efficient investment portfolio. In R. Silhavy, P. Silhavy, & Z. Prokopova (Eds.), *Software engineering perspectives in intelligent systems. CoMeSySo 2020. Advances in intelligent systems and computing* (Vol. 1294). Springer. https://doi.org/10.1007/978-3-030-63322-6_39

Ivanyuk, V., & Levchenko, K. (2020). Intelligent methods for predicting financial time series. In R. Silhavy, P. Silhavy, & Z. Prokopova (Eds.), *Software engineering perspectives in intelligent systems. CoMeSySo 2020. Advances in intelligent systems and computing* (Vol. 1294). Springer. https://doi.org/10.1007/978-3-030-63322-6_41

Ivanyuk, V., & Soloviev, V. (2019). Efficiency of neural networks in forecasting problems. In *Proceedings of 2019 12th International Conference Management of Large-Scale System Development, MLSD 2019*. https://doi.org/10.1109/MLSD.2019.8911046

Ivanyuk, V., Sunchalin, A., & Sunchalina, A. (2020). Development of an intelligent ensemble forecasting system. In R. Silhavy, P. Silhavy, & Z. Prokopova (Eds.), *Software Engineering Perspectives in Intelligent Systems. CoMeSySo 2020. Advances in intelligent systems and computing* (Vol. 1294). Springer. https://doi.org/10.1007/978-3-030-63322-6_40

Li, X., Zhu, S., Yüksel, S., Dinçer, H., & Ubay, G. G. (2020). Kano-based mapping of innovation strategies for renewable energy alternatives using hybrid interval type-2 fuzzy decision-making approach. *Energy, 211*, 118679.

Lisin, A. (2020). Valuation of the activities of foreign banks in the Russian banking sector. *Orbis, 15*(45), 53–63. http://www.revistaorbis.org.ve/pdf/45/art5.pdf

Liu, Z.-J., Panfilova, E., Mikhaylov, A., & Kurilova, A. (2021). COVID-19 crisis impact on the stability between parties in crowdfunding and crowdsourcing. *Wireless Personal Communications*. https://doi.org/10.1007/s11277-021-08932-z

Liu, Z.-J., Panfilova, E., Mikhaylov, A., & Kurilova, A. (2022). Assessing stability in relationship between parties in crowdfunding and crowdsourcing projects during the COVID-19 crisis. *Journal of Global Information Management, 30*(2), 1.

Melnichuk, A. V., Melnichuk, Y. A., Bocharov, A. V., Stepanova, D. I., & Kurbanov, S. A. (2020). Constitutional foundations of the functioning of local self-government. *Revista Inclusiones, 7*, 337–346.

Meynkhard, A. (2019). Fair market value of bitcoin: Halving effect. *Investment Management and Financial Innovations, 16*(4), 72–85. https://doi.org/10.21511/imfi.16(4).2019.07

Meynkhard, A. (2020). Effect of bitcoin volatility on altcoins pricing. *Advances in Intelligent Systems and Computing, 1294*, 652–664. https://doi.org/10.1007/978-3-030-63322-6_55

Mikhaylov, A. (2018a). Volatility spillover effect between stock and exchange rate in oil exporting countries. *International Journal of Energy Economics and Policy, 8*(3), 321–326.

Mikhaylov, A. (2018b). Pricing in oil market and using probit model for analysis of stock market effects. *International Journal of Energy Economics and Policy, 8*(2), 69–73.

Mikhaylov, A. (2018c). The yield of the carry trade strategy. *Finance: Theory and Practice, 22*(3), 52–63. https://doi.org/10.26794/2587-5671-2018-22-3-52-63

Mikhaylov, A. (2019). Oil and gas budget revenues in Russia after crisis in 2015. *International Journal of Energy Economics and Policy, 9*(2), 375–380.

Mikhaylov, A. (2020a). Cryptocurrency market analysis from the open innovation perspective. *Journal of Open Innovation: Technology, Market, and Complexity, 6*(4), 197. https://doi.org/10.3390/joitmc6040197

Mikhaylov, A. (2020b). Geothermal energy development in Iceland. *International Journal of Energy Economics and Policy, 10*(4), 31–35.

Mikhaylov, A. (2020c). Cryptocurrency market development: Hurst method. *Finance: Theory and Practice, 24*(3), 81–91.

Mikhaylov, A. (2021a). Development of Friedrich von Hayek's theory of private money and economic implications for digital currencies. *Terra Economicus, 19*(1), 53–62.

Mikhaylov, A. (2021b). Lichens as indicators of atmospheric pollution in urban ecosystems. *Israel Journal of Ecology & Evolution, 67*(1–2), 60–68. https://doi.org/10.1163/22244662-bja10016

Mikhaylov, A. (2022a). Efficiency of renewable energy generation in Russia. *Annals of the Brazilian Academy of Sciences, 94*(3).

Mikhaylov, A. (2022b). Factors of financial development and open innovation in Russia. *Journal of Open Innovation: Technology, Market, and Complexity, 8*(1), 197.

Mikhaylov, A., Burakov, D., & Didenko, V. (2019). Relationship between oil Price and macroeconomic indicators in Russia. *Finance: Theory and Practice, 23*(2), 105–116. https://doi.org/10.26794/2587-5671-2019-23-2-105-116

Mikhaylov, A., Danish, M. S. S., & Senjyu, T. (2021b). New stage in evolution of cryptocurrency market: Analysis by Hurst method. In H. Dincer & S. Yuksel (Eds.), *Strategic outlook in business and finance innovation: Multidimensional policies for emerging economies* (pp. 35–45. ISBN: 978-1-80043-445-5). Emerald Publishing Limited. https://doi.org/10.1108/978-1-80043-444-820211004

Mikhaylov, A., Moiseev, N., Aleshin, K., & Burkhardt, T. (2020). Global climate change and greenhouse effect. *Entrepreneurship and Sustainability Issues, 7*(4), 2897–2913.

Mikhaylov, A., & Sokolinskaya, N. (2019). Russian banks after sanctions of 2014. *Orbis, 15*(44), 55–65.

Mikhaylov, A., Sokolinskaya, N., & Nyangarika, A. (2018). Optimal carry trade strategy based on currencies of energy and developed economies. *Journal of Reviews on Global Economics, 7*, 582–592.

Mikhaylov, A., & Tarakanov, S. (2020). Development of Levenberg-Marquardt theoretical approach for electric network. *Journal of Physics: Conference Series, 1515*, 052006.

Mikhaylov, A., Yumashev, A., & Kolpak, E. (2021a). Quality of life, anxiety and depressive disorders in patients with extrasystolic arrhythmia. *Archives of Medical Science, 17*(1). https://doi.org/10.5114/aoms.2020.101359

Moiseev, N., Mikhaylov, A., Varyash, I., & Saqib, A. (2020). Investigating the relation of GDP per capita and corruption index. *Entrepreneurship and Sustainability Issues, 8*(1), 780–794.

Moiseev, N., Sorokin, A., Zvezdina, N., Mikhaylov, A., Khomyakova, L., & Danish, M. S. S. (2021). Credit risk theoretical model on the base of DCC-GARCH in time-varying parameters framework. *Mathematics, 9*, 2423. https://doi.org/10.3390/math9192423

Morkovkin, D. E., Gibadullin, A. A., Kolosova, E. V., Semkina, N. S., & Fasehzoda, I. S. (2020a). Modern transformation of the production base in the conditions of industry 4.0: Problems and prospects. *Journal of Physics: Conference Series, 1515*, 032014.

Morkovkin, D., Lopatkin, D., Sadriddinov, M., Shushunova, T., Gibadullin, A., & Golikova, O. (2020b). Assessment of innovation activity in the countries of the world. *E3S Web of Conferences, 157*, 04015. https://doi.org/10.1051/e3sconf/202015704015

Mukhametov, A., Bekhorashvili, A., Avdeenko, A., & Mikhaylov, A. (2021). The impact of growing legume plants under conditions of Biologization and soil cultivation on Chernozem fertility and productivity of rotation crops. *Legume Research, 44*, 12.

Mutalimov, V., Kovaleva, I., Mikhaylov, A., & Stepanova, D. (2021). Assessing regional growth of small business in Russia. *Entrepreneurial Business and Economics Review, 9*(3), 119–133. https://doi.org/10.15678/EBER.2021.090308

Nie, D., Panfilova, E., Samusenkov, V., & Mikhaylov, A. (2020). E-learning financing models in Russia for sustainable development. *Sustainability, 12*(11), 4412.

Nyangarika, A., Mikhaylov, A., & Richter, U. (2019a). Influence oil price towards economic indicators in Russia. *International Journal of Energy Economics and Policy, 9*(1), 123–129.

Nyangarika, A., Mikhaylov, A., & Richter, U. (2019b). Oil price factors: Forecasting on the base of modified auto-regressive integrated moving average model. *International Journal of Energy Economics and Policy, 9*(1), 149–159.

Nyangarika, A., Mikhaylov, A., & Tang, B.-J. (2018). Correlation of oil prices and gross domestic product in oil producing countries. *International Journal of Energy Economics and Policy, 8*(5), 42–48.

Qiu, D., Dinçer, H., Yüksel, S., & Ubay, G. G. (2020). Multi-faceted analysis of systematic risk-based wind energy investment decisions in E7 economies using modified hybrid modeling with IT2 fuzzy sets. *Energies, 13*(6), 1423.

Radosteva, M., Soloviev, V., Ivanyuk, V., & Tsvirkun, A. (2018). Use of neural network models in market risk management. *Advances in Systems Science and Applications, 18*(2), 53–58. https://doi.org/10.25728/assa.2018.18.2.582

Ranjbar, O., Chang, T., Nel, E., & Gupta, R. (2017). Energy consumption and economic growth nexus in South Africa: Asymmetric frequency domain approach. *Energy Sources, Part B: Economics, Planning, and Policy, 12*(1), 24–31.

Rathnayaka, R. K. T., Seneviratna, D. M. K., & Long, W. (2018). The dynamic relationship between energy consumption and economic growth in China. *Energy Sources, Part B: Economics, Planning, and Policy, 13*(5), 264–268.

Saqib, A., Chan, T.-H., Mikhaylov, A., & Lean, H. H. (2021). Are the responses of sectoral energy imports asymmetric to exchange rate volatilities in Pakistan? Evidence from recent foreign exchange regime. *Frontiers in Energy Research, 9*, 614463. https://doi.org/10.3389/fenrg.2021.614463

Sediqi, M. M., Nakadomari, A., Mikhaylov, A., Krishnan, N., Lotfy, M. E., Yona, A., & Senjyu, T. (2022). Impact of time of-use demand response program on optimal operation of Afghanistan real power system. *Energies, 15*, 296. https://doi.org/10.3390/en15010296

Sunchalin, A. M., Kochkarov, R. A., Levchenko, K. G., Kochkarov, A. A., & Ivanyuk, V. A. (2019). Methods of risk management in portfolio theory. *Espacios, 40*(16), 8.

Uandykova, M., Lisin, A., Stepanova, D., Baitenova, L., Mutaliyeva, L., Yuksel, S., & Dincer, H. (2020). The social and legislative principles of counteracting ransomware crime. *Entrepreneurship and Sustainability Issues, 8*(2), 777–798.

Udalov, I. (2021). The transition to renewable energy sources as a threat to resource economies. *International Journal of Energy Economics and Policy, 11*(3), 1–6.

Uyeh, D. D., Asem-Hiablie, S., Park, T., Kim, K.-M., Mikhaylov, A., Woo, S., & Ha, Y. (2021). Could japonica Rice be an alternative variety for increased global food security and climate change mitigation? *Food, 10*, 1869. https://doi.org/10.3390/foods10081869

Varyash, I., Mikhaylov, A., Moiseev, N., & Aleshin, K. (2020). Triple bottom line and corporate social responsibility performance indicators for Russian companies. *Entrepreneurship and Sustainability Issues, 8*(1), 313–331.

Wang, S., Liu, Q., Yuksel, S., & Dincer, H. (2019). Hesitant linguistic term sets-based hybrid analysis for renewable energy investments. *IEEE Access, 7*, 114223–114235.

Yuan, G., Xie, F., Dinçer, H., & Yüksel, S. (2021). The theory of inventive problem solving (TRIZ)-based strategic mapping of green nuclear energy investments with spherical fuzzy group decision-making approach. *International Journal of Energy Research*. https://doi.org/10.1002/er.6435

Yüksel, S., Dinçer, H., Çağlayan, Ç., & Uluer, G. S. (2021a). Determining the optimal financial strategies for nuclear energy companies. In *Financial strategies in competitive markets* (pp. 1–16). Springer.

Yüksel, S., Khomyakova, L., & Mikhaylov, A. (2021b). Energy Center selection in G7 industry with fuzzy MOORA. In S. Yüksel & H. Dinçer (Eds.), *Handbook of research on strategic Management for Current Energy Investments* (pp. 87–106). IGI Global. ISBN13: 9781799883357. https://doi.org/10.4018/978-1-7998-8335-7.ch006

Yüksel, S., Mikhaylov, A., & Ubay, G. G. (2021d). Factors causing delay in the installation of nuclear power plants. In S. Yüksel & H. Dinçer (Eds.), *Strategic approaches to energy management. Contributions to management science*. Springer. https://doi.org/10.1007/978-3-030-76783-9_7

Yüksel, S., Mikhaylov, A., Ubay, G. G., & Uyeh, D. D. (2021c). The role of hydrogen in the Black Sea for the future energy supply security of Turkey. In S. Yüksel & H. Dinçer (Eds.), *Handbook of research on strategic Management for Current Energy Investments* (pp. 1–15). IGI Global. https://doi.org/10.4018/978-1-7998-8335-7.ch001

Yumashev, A., & Mikhaylov, A. (2020). Development of polymer film coatings with high adhesion to steel alloys and high wear resistance. *Polymer Composites, 41*(7), 2875–2880.

Yumashev, A., Ślusarczyk, B., Kondrashev, S., & Mikhaylov, A. (2020). Global indicators of sustainable development: Evaluation of the influence of the human development index on consumption and quality of energy. *Energies, 13*, 2768.

Zhao, Q., Cherkasov, S., Avdeenko, A., Kondratenko, L., & Mikhaylov, A. (2021). Integral estimate of the added value of agricultural crops in the synergetic agriculture on the example of vicia faba (Vicia faba L.). *Polish Journal of Environmental Studies, 30*(5), 1–9. https://doi.org/10.15244/pjoes/131949

Zhou, P., Luo, J., Cheng, F., Yüksel, S., & Dinçer, H. (2021). Analysis of risk priorities for renewable energy investment projects using a hybrid IT2 hesitant fuzzy decision-making approach with alpha cuts. *Energy, 224*, 120184.

Development, Trade Openness, and Pollution: Is there any Threshold?

Fatma Taşdemir

1 Introduction

The Environmental Kuznets Curve (EKC) posits an inverted-U shaped relationship between the pollution and real income per capita. Accordingly, pollution increases with income up to a certain threshold level beyond which environment improves with income. In the early stage of development, people demand "more job" than clean environment (Dasgupta et al., 2002) because the economies are "too poor to be green" (Martinez-Alier, 1995). In higher income levels, more demand for clean environment leads the countries to follow "green growth" path. Therefore, the inverted-U shaped EKC suggests that environmental quality first worsens and then improves with the income level.

Grossman and Krueger (1993) explains the inverted-U shaped relation by scale, composition, and technique effects. The scale effect denotes the higher environmental degradation caused by higher income. The composition and technique effects mainly refer to the environmental enhancement due to the transition of economic structure from emission intensive industrial sector to the greener services sector and employment of technologies that aim to reduce pollution. The literature also provides several reasons that explain the inverted-U shaped relation between the pollution measure and income per capita (Kijima et al., 2010; Dasgupta et al., 2002; Dinda, 2004). First, the demand for clean environment matters more for the countries reaching a certain threshold level of income. Second, in line with the composition effect argument by Grossman and Krueger (1993), the reallocation of sectors from agriculture to pollution intensive industry and from industry to "green" services has been led to first increasing then decreasing relationship between

F. Taşdemir (✉)
Sinop University, Sinop, Turkey
e-mail: ftasdemir@sinop.edu.tr

© The Author(s), under exclusive license to Springer Nature Switzerland AG 2022
H. Dinçer, S. Yüksel (eds.), *Sustainability in Energy Business and Finance*, Contributions to Finance and Accounting,
https://doi.org/10.1007/978-3-030-94051-5_10

pollution and income. Third, in accord with the technique effect explanation by Grossman and Krueger (1993), high income levels enable the countries to prioritize the environment friendly technologies by investing more on research and development activities. Fourth, economies with better institutional quality and governance are more likely to implement environment friendly policies than the others.

Dinda (2004) and Shahbaz and Sinha (2019) provide a brief literature review on the EKC. The bulk of the literature often investigates the validity and turning point of the inverted-U shaped relation. The empirical findings by Millimet et al. (2003) suggest the validity of inverted-U shaped relation while Stern (2004) indicates that the relationship between pollution and income is neither universal nor robust. Aslanidis and Iranzo (2009) and Şentürk et al. (2020) find that there is a monotonically increasing relation between the pollution and income, albeit this positive relationship is much lower in high income levels. The estimation results based on panel smooth transition regression by Taşdemir (2021) suggest that the sensitivity of pollution to income is substantially much higher in economies with higher manufacturing and industry sectors share in GDP while it is much lower in countries with higher services sector share in income. In the context of estimating the turning point of the inverted-U shaped relation, Kaika and Zervas (2013) maintain that turning point of the EKC may not necessarily be the same in advanced economies with people earn above the world average income than the others. The empirical findings by Churchill et al. (2018) suggest that turning point of the EKC in OECD countries is between the \$18,955 and \$89,540 per capita income levels. On the other hand, the results by Martínez-Zarzoso and Bengochea-Morancho (2004) suggest turning point of the inverted-U shaped curve in OECD economies is between \$4914 and \$18,364 per capita income levels. The empirical findings by Roberts and Grimes (1997) suggest that the EKC appears to be hold in high income economies due to their use of energy-efficient technologies. The results by Galeotti et al. (2006) suggest that the shape of the EKC changes depending on the data sources. In a similar vein, the findings by Harbaugh et al. (2002) indicate the inverted-U shaped relation may change depending on the employment of alternative pollution measures.

The literature also explains the CO_2 emissions-income relation by the pollution haven hypothesis (PHH) that considers the environmental effect of trade and foreign direct investments (FDI). Porter (1999) suggests the “race to the bottom” and “stuck at the bottom” explanations to investigate the environmental effects of trade. Accordingly, “stuck at the bottom” mainly refers to the lax environmental standards prevailed in industrializing countries lead them to have comparative advantage in the production of pollution intensive goods. On the other hand, “race to the bottom” denotes the unregulated trade may cause the loosening in environmental standards in industrialized countries with well-designed environmental regulations. In the context of trade and pollution, Porter (1999) suggests an arrangement that requires minimum standards in environmental regulation, especially in industrializing countries. According to Antweiler et al. (2001), trade openness leads to a decrease in emissions due to the efficient use of resources while Andreoni and Levinson (2001) find that trade openness causes developed countries to locate dirty industries to developing countries with lower environmental standards. The findings by Dean (2002) suggest that the impact of trade liberalization can both lead to environmental degradation if there is an improvement in the terms of trade and environmental enhancement if trade openness promotes growth. The short- and long-run effect of trade openness is associated with higher pollution in non-OECD economies while lower pollution in OECD economies according to the results by Managi et al. (2009). Kearsley and Riddel (2010) and Frankel and Rose (2005) suggest that there is a weak relationship between trade and pollution. Aklin (2016) finds that trade plays a crucially important role in the emission reduction that observed in industrialized countries. Shapiro and Walker (2018) find that the rise in implicit pollution tax in the U.S. than trade has been led to a substantial decrease in emissions from the manufacturing sector.

The seminal studies by Grossman and Krueger (1993, 1995) have been led to the investigation of the causes of pollution. In line with the above-reviewed literature, we maintain that income and trade openness are important determinants of global pollution measured by CO_2 emissions. Kearsley and Riddel (2010, p. 905) notes that if the pollution haven hypothesis is valid, then omitting the trade openness "may bias the estimate of the EKC's turning point." The bulk of the literature often investigates the validity of inverted-U shaped EKC by regressing the pollution measure on income per capita and squared income per capita. Lind and Mehlum (2010) suggest that the empirical findings with the positively significant coefficient for income per capita and negatively significant coefficient for squared income per capita provide only a weak support to the validity of inverted-U shaped relationship. Şentürk et al. (2020, p. 5) note that the conventional EKC based on often quadratic regression models "may not identify other forms of non-linearity that may exist." Therefore, the main aim of this chapter is to investigate the relationships between income per capita, trade openness, and global pollution measure of CO_2 emissions in 87 countries over the 1970–2019 period. Considering the potential endogeneity, we prefer to use dynamic panel threshold procedure by Kremer et al. (2013). First, for a given trade openness, we investigate whether the impact of income per capita on pollution may change depending on the income levels of the economies. Then, for a given income per capita, we examine whether the sensitivity of CO_2 emissions to trade openness may change with respect to the trade openness levels.

This study finds that, for a given trade openness level, income per capita provides data-driven threshold for the impact of income per capita on CO_2 emissions. This threshold level is around \$25,000 for the sample of advanced economies. An increase in income lowers the pollution and the income elasticity of pollution is almost the same in advanced economies with low- and high-income levels. On the other hand, the threshold level of income per capita is almost \$3900 in emerging market and developing economies. An increase in income leads to higher pollution, albeit the income elasticity of pollution is slightly lower in emerging market and developing economies with higher income. We find that, for a given income per capita, the level of trade openness also constitutes endogenous threshold in explaining the effect of trade on pollution. The threshold level of trade openness (as a percent of GDP) is around 190 in advanced economies. Trade openness leads to lower pollution, albeit this is slightly higher in economies with less trade integrated.

The threshold level of trade openness (as a percent of GDP) is almost 110 in emerging market and developing economies. Trade openness leads to an increase in environmental degradation in more trade open economies. The empirical findings in this study suggest that there is a monotonically decreasing and increasing relationships between income and pollution, respectively, in advanced and emerging market and developing economies. The pollution haven hypothesis appears to be hold in emerging market and developing economies. This may imply that advanced economies may locate some pollution intensive productions to lax environmental regulations prevailed emerging market and developing economies by trade linkages.

The plan for the rest of this chapter is as follows. Section 2 presents the empirical methodology and reports the estimation results. Section 2.1 considers the thresholding effect of income per capita. Section 2.2 investigates the thresholding effect of trade openness. Section 3 concludes and provides some policy implications.

2 Empirical Methodology and Estimation Results

To investigate the relationship between CO_2 emissions, income per capita, and trade openness, we consider the following benchmark equation:

$$CO_{2,it} = \alpha_i + \alpha_1 CO_{2,it-1} + \alpha_2 GDPpc_{it} + \alpha_3 TRADE_{it} + u_{it} \tag{1}$$

In Eq. (1), the subscript i and t represent, respectively, country and time; CO_2 is the natural logarithm of CO_2 emissions per tones per capita, GDPpc is the natural logarithm of real GDP per capita (in constant 2015 US dollars), and TRADE is the trade openness measured by the sum of exports and imports of goods and services as a percent of GDP. We include the lagged CO_2 because the current level of emissions may also depend on the past values. The main data source for CO_2 emissions is Joint Research Centre Emissions Database for Global Atmospheric Research. Real GDP per capita data are from United Nations Conference on Trade and Development database. The data for trade openness are from World Development Indicators, World Bank.

Equation (1) maintains that CO_2 emissions can be explained mainly with income per capita and trade openness. This equation suggests that, given the level of trade openness, the income elasticity of CO_2 emissions is invariant to the income levels. However, per capita income levels may provide data-driven threshold for the income elasticity of CO_2 emissions. This may also be the case for trade openness. Equation (1) suggests that, given the level of income per capita, the impact of trade openness on CO_2 emissions is the same in economies with low and high trade integrated. Trade openness may also provide endogenously estimated threshold in explaining the effect of trade on CO_2 emissions.

The conventional EKC literature often considers the nonlinearity and/or threshold issues either by employing quadratic or cubic regression models. Lind and Mehlum (2010) suggest that the calculation of turning points based on the quadratic regression models may provide only a weak support to the validity of inverted-U shaped relationship between per capita income and pollution measure. Furthermore, they maintain that this may inadvertently yield an inverted U-shaped EKC "when the true relationship is convex but monotone over the relevant data values" (p. 110). Also, Şentürk et al. (2020, p. 5) suggest that the consideration of nonlinearity based on the quadratic or cubic regression models "may not identify other forms of non-linearity that may exist." In this context, the main aim of this chapter is to investigate the potential thresholding effects of income per capita and trade openness based on the data-driven estimated threshold procedures. We investigate this crucially important issue empirically for an unbalanced panel of 87 advanced[1] and emerging market and developing[2] economies over the 1970–2019 period by employing dynamic panel threshold procedure of Kremer et al. (2013).

Considering the potential endogeneity concerns, we investigate the relationship between CO_2 emissions, income per capita, and trade openness by employing dynamic panel threshold procedure. As suggested by Kremer et al. (2013), the first step of the estimation includes elimination of country-specific fixed effects via forward orthogonal transformation that removes the serial autocorrelation concerns. We regressed the endogenous variable, i.e. lagged CO_2 emissions on a set of instruments consisting of the higher lags of CO_2 emissions. Then, by substituting the predicted values of lagged CO_2 emissions into the benchmark equation, we employ panel least squares estimation procedure to estimate the separate thresholding effects of income per capita and trade openness. By employing Hansen (1999) procedure, we first estimate the threshold that yields the minimum sum of squared residuals. After finding a statistically significant thresholding effect of income per capita and trade openness, we employ the generalized method of moments (GMM) procedure to estimate the slope parameters.

[1] Advanced economies sample consists of Australia, Austria, Belgium, Canada, Cyprus, Denmark, Finland, France, Germany, Greece, Hong Kong, Iceland, Ireland, Italy, Japan, Luxembourg, Malta, Netherlands, New Zealand, Norway, Portugal, Singapore, Spain, Sweden, Switzerland, United Kingdom, and United States.

[2] The sample of emerging market and developing economies includes Albania, Argentina, Bangladesh, Belize, Benin, Bhutan, Bolivia, Botswana, Brazil, Bulgaria, Cabo Verde, Cambodia, Cameroon, Chile, China, Colombia, Costa Rica, Cote d'Ivoire, Dominican Republic, Ecuador, Egypt, El Salvador, Fiji, Ghana, Guatemala, Guyana, Haiti, Honduras, Hungary, India, Indonesia, Israel, Jamaica, Jordan, Kenya, Korea Republic, Lebanon, Malaysia, Mauritania, Mauritius, Mexico, Morocco, Nepal, Pakistan, Panama, Papua New Guinea, Paraguay, Peru, Philippines, Poland, Romania, Senegal, South Africa, Sri Lanka, Thailand, Tonga, Tunisia, Turkey, Uruguay, and Zimbabwe.

Table 1 Thresholding effect of income per capita

	Whole sample	Advanced economies	Emerging market and developing economies
Threshold $GDPpc_{it}$	9.989	10.134	8.279
95% CI	[9.988, 10.385]	[9.941, 10.689]	[8.278, 8.846]
Antilog (GDPpc*)	\$21,778	\$25,197	\$3940
NT_{TH}	1092	978	1143
$GDPpc_{it}$ ($GDPpc_{it} \leq \lambda$)	0.087*** (0.020)	−0.020** (0.010)	0.140*** (0.026)
$GDPpc_{it}$ ($GDPpc_{it} > \lambda$)	0.072*** (0.017)	−0.019** (0.010)	0.133*** (0.024)
$TRADE_{it}$	−0.046** (0.019)	−0.026** (0.013)	0.061** (0.025)
$CO_{2i,\ t-1}$	0.928*** (0.014)	0.951*** (0.011)	0.881*** (0.017)
Constant	−0.607*** (0.154)	0.332*** (0.097)	−1.091*** (0.203)
NT	4004	1310	2694
N	87	27	60

Notes: The values in parentheses are the standard errors. *, **, and ***, respectively, denote significance at 10%, 5%, and 1% levels. *N* and NT are, correspondingly, the numbers of countries and the effective number of observations. NT_{TH} shows the number of observations above the estimated data-driven threshold level

3 Thresholding Effect of Income

To investigate whether income per capita (GDPpc) provides data-driven threshold for the impact of per capita income on CO_2 emissions, we consider the following dynamic equation:

$$CO_{2,it} = \alpha_i + \alpha_1 CO_{2,it-1} + \alpha_2 GDPpc_{it}(GDPpc_{it} \leq \lambda) + \alpha_3 GDPpc_{it}(GDPpc_{it} > \lambda) + \alpha_4\ TRADE_{it} + u1_{it} \tag{2}$$

In Eq. (2), λ is data-driven estimated threshold. This threshold divides the observations in the sample as low and high regimes. For instance, if income per capita is higher than the endogenously determined threshold ($GDPpc_{it} > \lambda$), then the estimated coefficient α_3 shows the impact of income per capita on CO_2 emissions in the high regime. Otherwise, the estimated coefficient (α_2) shows the effect of income per capita on pollution in the low regime. Thus, the low and high regimes are differentiated from each other with different slope parameters.

Table 1 reports the dynamic panel threshold estimation results of Eq. (2). The endogenously determined threshold value is almost 10 for the whole sample. As reported by the Table, this threshold value is included within the 95% confidence interval. This may suggest that income per capita provides a significant threshold effect. The data-driven estimated threshold value corresponds to $21,778 per capita income level which is slightly higher than the mean that is around $13,000. The high regime contains almost 30% of the observations. The income elasticity of pollution is 0.09 in economies with per capita income is less than $21,778 while it is around 0.07 in countries with per capita income is higher than $21,778. This empirical finding may suggest that income per capita leads to higher CO_2 emissions, albeit this impact is slightly lower in high income economies. This may also suggest that there is a monotonically increasing relation between income per capita and pollution in contrast to the maintained inverted-U shaped EKC. According to the results, trade openness (TRADE) appears to provide an improvement in environmental quality. The estimated coefficient for lagged emission may indicate that countries with high per capita CO_2 emissions tend to experience a faster decrease in emissions.

The data-driven estimated threshold is almost 10.1 for the advanced economies. The endogenously determined threshold value is contained within the 95% confidence interval. This provides an empirical support to the significant thresholding effect of income per capita. The endogenously determined threshold value corresponds to $25,197 per capita income level with almost 75% of the observations are in the high regime. This threshold is slightly lower than the mean which is almost $34,000. The income elasticity of CO_2 emissions is negative and statistically significant suggesting higher income increases the environmental quality, albeit the estimated coefficient for income per capita is almost the same in both regimes. This may also suggest that there is a monotonically decreasing relationship between income and pollution in advanced economies. An increase in trade openness leads to lower CO_2 emissions. The coefficient for lagged CO_2 emissions suggests that ecological convergence appears to be hold in advanced economies.

The endogenously determined threshold value of income per capita is around 8.3 for emerging market and developing economies. The data-driven estimated threshold lies within the 95% confidence interval suggesting the presence of significant thresholding effect of income per capita. The threshold value is almost the same with the mean and corresponds to $3940 per capita income level with almost 45% of the observations are in the high regime. As compared to advanced economies, the threshold level of income per capita is substantially much lower in emerging market and developing economies. The income elasticity of CO_2 emissions is 0.14 in economies with per capita income level is lower than $3940 while it is around 0.13 in economies with per capita income level is higher than $3940. An increase in per capita income leads to higher CO_2 emissions, albeit the income elasticity of pollution is slightly lower in the high regime. In contrast to the postulated inverted-U shaped EKC, this empirical finding may suggest that there is a monotonically increasing relation between income and pollution. Higher trade openness increases CO_2 emissions. The estimated coefficient for lagged CO_2 emissions indicates that countries with high per capita CO_2 emissions tend to experience a faster decrease in pollution.

4 Thresholding Effect of Trade Openness

To investigate whether trade openness (TRADE) provides data-driven threshold for the impact of trade on CO_2 emissions, we consider the following dynamic equation:

$$CO_{2,it} = \alpha_i + \alpha_1 CO_{2,it-1} + \alpha_2 \text{TRADE}_{it}(\text{TRADE}_{it} \leq \lambda) + \alpha_3 \text{TRADE}_{it}(\text{TRADE}_{it} \leq \lambda) + \alpha_4\, \text{GDPpc}_{it} + u2_{it} \tag{3}$$

Table 2 reports the dynamic panel threshold estimation results for Eq. (3). Accordingly, trade openness provides data-driven threshold in explaining the impact of trade on CO_2 emissions. The mean of trade openness (as a percent of GDP) is around 74 for the whole sample. The endogenously determined threshold value of trade is 81 which is almost the same with the mean and around 35% of the observations are in the high regime. Trade openness leads to a decrease in CO_2 emissions in the high regime (TRADE >81.22) consisting of more trade open

Table 2 Thresholding effect of trade openness

	Whole sample	Advanced economies	Emerging market and developing economies
Threshold TRADE_{it}	81.22	191.55	115.71
95% CI	[41.83; 100.05]	[51.98; 210.58]	[109.46; 116.55]
NT_{TH}	1384	156	291
TRADE_{it} $(\text{TRADE}_{it} \leq \lambda)$	−0.017 (0.024)	−0.042*** (0.013)	0.018 (0.026)
TRADE_{it} $(\text{TRADE}_{it} > \lambda)$	−0.050*** (0.015)	−0.030*** (0.009)	0.054** (0.021)
GDPpc_{it}	0.045*** (0.010)	−0.012* (0.007)	0.113*** (0.014)
$CO_{2i,\,t-1}$	0.944*** (0.008)	0.950*** (0.009)	0.885*** (0.011)
Constant	−0.309*** (0.080)	0.265*** (0.073)	−0.876*** (0.110)
NT	3942	1254	2688
N	87	27	60

Notes: The values in parentheses are the standard errors. *, **, and ***, respectively, denote significance at 10%, 5%, and 1% levels. *N* and NT are, correspondingly, the numbers of countries and the effective number of observations. NT_{TH} shows the number of observations above the estimated data-driven threshold level economies. This finding may suggest that there is a certain threshold level of trade openness to reap the benefits in terms of environmental enhancement. The income elasticity of pollution is positive and statistically significant suggesting higher income increases the pollution. There is an ecological convergence in per capita CO_2 emissions for the whole sample.

The data-driven estimated threshold of trade openness is around 192 for the sample of advanced economies with almost 13% of the observations are in the high regime. As compared to the mean of trade, which is around 97, this threshold value may be interpreted as slightly high. Trade openness provides an improvement in environmental quality in both regimes. This is mainly in line with the empirical findings by Managi et al. (2009) and Aklin (2016) indicating that industrialized countries mainly consist of advanced economies with high environmental standards may locate dirty industries to lax environmental regulations prevailed developing economies by trade linkages. The income elasticity of pollution is negative and statistically significant suggesting that an increase in income leads to a decrease in CO_2 emissions. The estimated coefficient for lagged pollution suggests that advanced economies converge to each other in terms of per capita CO_2 emissions.

The endogenously determined threshold value of trade is almost 115 in emerging market and developing economies with around 11% of the observations are in the high regime. Considering the mean of trade is 64, the data-driven estimated threshold may be interpreted as slightly high. Trade openness leads to an increase in CO_2 emissions in more trade open economies. This finding is consistent with the "stuck at the bottom" explanation provided by Porter (1999) suggesting that industrializing countries mainly consist of emerging market and developing economies may have lax environmental standards leading them to have a comparative advantage in the production of pollution intensive goods. The income elasticity of pollution is positive and statistically significant suggesting that an increase in income rises pollution. There is an ecological convergence in per capita CO_2 emissions for the sample of emerging market and developing economies.

5 Conclusion

The conventional environmental Kuznets curve (EKC) posits an inverted-U shaped relationship between pollution and income per capita that represents the aggregate measure of economic activities. Accordingly, income leads to pollution up to a certain threshold beyond which income leads to improvement in environment. Therefore, environmental economists often suggest that income is both the cause and cure for environmental pollution. The bulk of the literature suggests that trade openness is also one of the most important determinants of environmental degradation. In this context, the main aim of this study is to investigate the relationship between income per capita, trade openness, and pollution in 87 advanced and emerging market and developing economies over the 1970–2019 period.

In the context of development, trade openness, and pollution, this study, first, investigates whether income per capita provides data-driven threshold for the impact of income per capita on pollution. Then, we examine whether the level of trade openness matters for the effect of trade on pollution. We investigate these crucially important research questions by employing dynamic panel threshold procedure of Kremer et al. (2013).

We find that, for a given level of trade openness, income per capita provides data-driven estimated threshold for the impact of income per capita on pollution. This threshold value is around \$25,000 for advanced economies and \$3900 for emerging market and developing economies. In income levels above and below the endogenously determined threshold, the income elasticity of pollution is negative and significant in advanced economies while it is positive and significant in emerging market and developing economies. In contrast to the maintained inverted-U shaped EKC, our estimation results suggest that the relationship between per capita income and pollution is monotonically decreasing in advanced economies while monotonically increasing in emerging market and developing economies. This empirical finding may suggest that the composition and technique effects explanations appear to be the case in advanced economies while the scale effect postulation appears to be hold in emerging market and developing economies.

We also find that, for a given level of income per capita, trade openness provides data-driven threshold for the impact of trade on pollution. The endogenously estimated threshold value for trade openness is around 190 in advanced economies while it is almost 110 in emerging market and developing economies. Trade openness leads to environmental improvement in advanced economies. On the other hand, trade openness increases pollution in more trade open emerging market and developing economies. In this context, Hermele (2002) notes that even if we live in a servicified economy represented by higher services sector share in GDP, we can benefit from the industrial goods produced elsewhere mainly by trade linkages. This is also consistent with the "stuck at the bottom" explanation provided by Porter (1999) suggesting that trade openness allows industrialized or servicified economies with strict environmental regulations to locate energy and pollution intensive industries into the industrializing countries with lax environmental standards.

The empirical findings in this study suggest that income and trade openness are important determinants of CO_2 emissions. The pulling and pushing impacts for better environment may lead the countries to invest in technologies that provide reduction in emissions. Given the recent rise in greenhouse gas emissions mostly sourced in CO_2 emissions, environmental problems may be considered as universal requiring cooperative solutions (Bhagwati, 1993). In this context, establishment of standardized environmental regulation to be able to trade has a strategic importance. This is much more important for emerging market and developing economies to reap the environmental benefits. Environment friendly proactive strategic management systems, incentivization of green investments, supporting the employment of alternative energy sources like solar and wind energies are all, indeed, crucially important policy suggestions that contribute to the sustainable development goals. All these indeed suggest that countries may better to design and enforce sustainable development policies by placing the greener economy at the core.

References

Aklin, M. (2016). Re-exploring the trade and environment nexus through the diffusion of pollution. *Environmental and Resource Economics, 64*(4), 663–682.

Andreoni, J., & Levinson, A. (2001). The simple analytics of the environmental Kuznets curve. *Journal of Public Economics, 80*, 269–286.

Antweiler, W., Copeland, B. R., & Taylor, M. S. (2001). Is free trade good for the environment? *American Economic Review, 91*(4), 877–908.

Aslanidis, N., & Iranzo, S. (2009). Environment and development: Is there a Kuznets curve for CO_2 emissions? *Applied Economics, 41*(6), 803–810.

Bhagwati, J. (1993). The case for free trade. *Scientific American, 269*(5), 42–49.

Churchill, S. A., Inekwe, J., Ivanovski, K., & Smyth, R. (2018). The environmental Kuznets curve in the OECD: 1870–2014. *Energy Economics, 75*, 389–399.

Dasgupta, S., Laplante, B., Wang, H., & Wheeler, D. (2002). Confronting the environmental Kuznets curve. *Journal of Economic Perspectives, 16*(1), 147–168.

Dean, J. M. (2002). Does trade liberalization harm the environment? A new test. *Canadian Journal of Economics/Revue canadienne d'économique, 35*(4), 819–842.

Dinda, S. (2004). Environmental Kuznets curve hypothesis: A survey. *Ecological Economics, 49*(4), 431–455.

Frankel, J. A., & Rose, A. K. (2005). Is trade good or bad for the environment? Sorting out the causality. *Review of Economics and Statistics, 87*(1), 85–91.

Galeotti, M., Lanza, A., & Pauli, F. (2006). Reassessing the environmental Kuznets curve for CO_2 emissions: A robustness exercise. *Ecological Economics, 57*(1), 152–163.

Grossman, G. M., & Krueger, A. B. (1993). Environmental impacts of the north American free trade agreement. In P. Garber (Ed.), *The U.S.–Mexico free trade agreement* (pp. 13–56). MIT Press.

Grossman, G. M., & Krueger, A. B. (1995). Economic growth and the environment. *The Quarterly Journal of Economics, 110*(2), 353–377.

Hansen, B. E. (1999). Threshold effects in non-dynamic panels: Estimation, testing, and inference. *Journal of Econometrics, 93*(2), 345–368.

Harbaugh, W. T., Levinson, A., & Wilson, D. M. (2002). Reexamining the empirical evidence for an environmental Kuznets curve. *Review of Economics and Statistics, 84*(3), 541–551.

Hermele, K. (2002). *Vad kostar framtiden? Globaliseringen, miljön och Sverige*. Ordfrontförlag.

Kaika, D., & Zervas, E. (2013). The environmental Kuznets curve (EKC) theory. *Part B: Critical issues. Energy Policy, 62*, 1403–1411.

Kearsley, A., & Riddel, M. (2010). A further inquiry into the pollution haven hypothesis and the environmental Kuznets curve. *Ecological Economics, 69*(4), 905–919.

Kijima, M., Nishide, K., & Ohyama, A. (2010). Economic models for the environmental Kuznets curve: A survey. *Journal of Economic Dynamics and Control, 34*(7), 1187–1201.

Kremer, S., Bick, A., & Nautz, D. (2013). Inflation and growth: New evidence from a dynamic panel threshold analysis. *Empirical Economics, 44*(2), 861–878.

Lind, J. T., & Mehlum, H. (2010). With or without U? The appropriate test for a U-shaped relationship. *Oxford Bulletin of Economics and Statistics, 72*(1), 109–118.

Managi, S., Hibiki, A., & Tsurumi, T. (2009). Does trade openness improve environmental quality? *Journal of Environmental Economics and Management, 58*(3), 346–363.

Martinez-Alier, J. (1995). The environment as a luxury good or "too poor to be green"? *Ecological Economics, 13*(1), 1–10.

Martínez-Zarzoso, I., & Bengochea-Morancho, A. (2004). Pooled mean group estimation of an environmental Kuznets curve for CO_2. *Economics Letters, 82*(1), 121–126.
Millimet, D. L., List, J. A., & Stengos, T. (2003). The environmental Kuznets curve: Real progress or misspecified models? *Review of Economics and Statistics, 85*(4), 1038–1047.
Porter, G. (1999). Trade competition and pollution standards:"race to the bottom" or "stuck at the bottom". *The Journal of Environment & Development, 8*(2), 133–151.
Roberts, J. T., & Grimes, P. E. (1997). Carbon intensity and economic development 1962–1991: A brief exploration of the environmental Kuznets curve. *World Development, 25*(2), 191–198.
Şentürk, H., Omay, T., Yildirim, J., & Köse, N. (2020). Environmental Kuznets curve: Non-linear panel regression analysis. *Environmental Modeling & Assessment, 25*(5), 633–651.
Shahbaz, M., & Sinha, A. (2019). Environmental Kuznets curve for CO_2 emissions: A literature survey. *Journal of Economic Studies, 46*(1), 106–168.
Shapiro, J. S., & Walker, R. (2018). Why is pollution from US manufacturing declining? The roles of environmental regulation, productivity, and trade. *American Economic Review, 108*(12), 3814–3854.
Stern, D. I. (2004). The rise and fall of the environmental Kuznets curve. *World Development, 32*(8), 1419–1439.
Taşdemir, F. (2021). Industrialization, servicification, and environmental Kuznets curve: Non-linear panel regression analysis. *Environmental Science and Pollution Research.* https://doi.org/10.1007/s11356-021-16012-9

Analysis of the Functioning of the Energy Safety Conditions

Diana Stepanova, Yulia Finogenova, Gabor Pinter, and Ismail Ismailov

1 Introduction

It is worth noting that in 2020, due to the depreciation of the ruble, the dynamics of deposits in nominal terms was quite good. The volume of energy risks increased by 8% (2.4 trillion rubles), which is not much different from the dynamics of recent years. Thus, almost half of the increase in nominal terms is a currency revaluation (Yüksel et al., 2021b). The low real growth rate in 2020 is due to the low profitability of energy risks, the rates on which are at a historically low level. Lower deposit rates have increased the public's interest in alternative methods of saving —investments in securities, mutual funds, life insurance, and others. Also, the promotion of mortgages and the record level of their issuance last year also limited the growth of deposits, as the population spent their savings in banks on the initial payment (Dinçer et al., 2020). It should be noted separately that during the pandemic and restrictions, the income of certain categories decreased citizens, which limited their ability to create savings. Against this background, the share of individual deposits in the banking system's liabilities decreased significantly, and as of January 1, 2021, it amounted to 31.6% against 34.2% as of January 1, 2020, and 33% as of January 1, 2019.

D. Stepanova (✉) · Y. Finogenova
Plekhanov Russian University of Economics, Russian Federation, Moscow, Russia
e-mail: Finogenova.YY@rea.ru

G. Pinter
Circular Economy University Center, Renewable Energy Research Group University of Pannonia, Veszprém, Hungary

I. Ismailov
Financial University under the Government of the Russian Federation, Russian Federation, Moscow, Russia
e-mail: iismailov@fa.ru

© The Author(s), under exclusive license to Springer Nature Switzerland AG 2022
H. Dinçer, S. Yüksel (eds.), *Sustainability in Energy Business and Finance*, Contributions to Finance and Accounting,
https://doi.org/10.1007/978-3-030-94051-5_11

We have prepared a rating of Russian banks by the volume of individual deposits as of January 1, 2021. The rating includes 320 banks that attract funds from the population, and for which reports are published in accordance with form No. 101 on the website of the Central bank of the Russian Federation. The rating methodology provides for aggregation of data from banks 'current statements. Of the credit institutions represented in the rating, 46.5% of banks in 2020 were able to demonstrate positive growth rates of attracted funds of individuals. For comparison, in 2017–2019, the share was more than 50%, and in 2016 and 2015, 62% and 76%, respectively. Thus, the coronavirus epidemic, low interest rates and falling revenues significantly affected the share of banks with an increase in deposits. In the quarterly context of the situation in 2020, the dynamics of the share of banks with growth was very high volatile. In particular, in the fourth quarter, deposits increased by only 39% compared to more than 50% in the first and third quarters. It should be noted separately that in the second quarter, which saw the main restrictions due to the coronavirus, the share of banks with an increase in deposits fell to 33%. Thus, the share of banks with an increase in deposits has not yet stabilized and is at a relatively low level.

2 Literature Review

Among different size groups of the TOP 300 banks in 2020, ranked by deposit size and divided by one hundred banks, the result fluctuated quite strongly. For example, 54% of banks out of the TOP 100 showed an increase in deposits. At the same time, 44% of banks in the second hundred (from 101 to 200 places) showed an increase in deposits. While 42% of banks ranked 201st to 300th had an increase in the population's energy risks. It should be noted separately that the largest credit institutions from the rating had an even higher share with growth, for example, the TOP 25 banks had an increase at 68%, and in the top ten, all banks were characterized by an increase in deposits (Mikhaylov, 2018c; Mikhaylov et al., 2019; Melnichuk et al., 2020; Nie et al., 2020; Moiseev et al., 2021; Qiu et al., 2020; Grilli et al., 2021).

The volume of energy risks controlled by the TOP 100 banks increased by 8% relative to the result of the same banks as of January 1, 2020. Banks of the second hundred (from 101 to 200 places) showed a decrease in total deposits by 3%. While the banking group from 201 to 300 places reduced deposits by a total of 6%. According to experts RIA Rating, the good result of large banks is due to the continued growth of concentration and, consequently, the flow of customers from small and medium-sized banks to large ones, as well as the fact that large banks are characterized by a greater share of currency energy risks (Bhuiyan et al., 2021; Dong et al., 2021; Mikhaylov, 2021b; Liu et al., 2022; Saqib et al., 2021; Zhou et al., 2021; Radosteva et al., 2018; Ranjbar et al., 2017; Rathnayaka et al., 2018; Sunchalin et al., 2019; Uandykova et al., 2020; Udalov, 2021; Yüksel et al., 2021d, 2021a, 2021c; Dorofeev, 2020).

The difference in the dynamics of individual deposits in different size groups led to a further increase in the concentration of household funds. The share of TOP-10 credit institutions in the energy risk market increased by 3.3 percentage points in 12 months to 80.8% as of January 1, 2021. The TOP 25 banks in the current rating already account for 90.4% of deposits (+1.8% in 12 months). At the same time, the share of the TOP 100 increased by 0.4 percentage points to 98.6% (Dayong et al., 2020; Mikhaylov et al., 2018; Nyangarika et al., 2018; Danish et al., 2020; Danish et al., 2021; Lisin, 2020; An et al., 2021; Ivanyuk & Berzin, 2020; Ivanyuk & Levchenko, 2020; Ivanyuk et al., 2020; Ivanyuk, 2018; Ivanyuk & Soloviev, 2019; Fang et al., 2021; Li et al., 2020; Du et al., 2020; Uyeh et al., 2021).

In 2020, among Russian banks, the best dynamics was demonstrated by banks with state participation. For 12 months of 2020, state-owned banks were able to nominally increase the amount of funds raised by the population by 8.6%. At the same time, private banks showed a slightly smaller increase—7.3%. At the same time, the result of banks with foreign control participation was +5.8% (Denisova et al., 2019; Nyangarika et al., 2019a, 2019b; Huang et al., 2021a, 2021b; Mikhaylov, 2018a, 2018b, 2022; Meynkhard, 2019; Conteh et al., 2021; Mikhaylov et al., 2019; Meynkhard, 2020).

The leader in the relative dynamics of deposits, among the TOP 100 banks in terms of energy risks, was a private bank—Energomashbank, which increased its deposit portfolio by more than two times in 2020. The second-largest bank in terms of growth among the hundred largest banks was the Bank Combined capital, the volume of energy risks, which increased by 56% in 12 months. Deposits also showed good dynamics in 2020 BANK ORENBURG and Lanta-Bank, the growth of deposits in which amounted to 46% for each of the banks (Mukhametov et al., 2021; Candila et al., 2021; Liu et al., 2021).

A significant amount of energy risks were also insured in 2020 VTB Bank (+283 billion rubles), Gazprombank (+236 billion rubles) and ALFA-BANK (+231 billion rubles). Thus, three state-owned banks provided just under 80% of the total increase in deposits of individuals in 2020. On the other hand, significant negative dynamics were demonstrated by: Bank Vozrozhdenie, HCF Bank and the Ministry of Finance, and the decrease in deposits from these banks amounted to 81, 46, and 45 billion rubles, respectively (An et al., 2019a, 2019b, 2020a, 2020b, 2020c; Mikhaylov et al., 2020; Mikhaylov & Tarakanov, 2020; Moiseev et al., 2020; Gura et al., 2020; Dooyum et al., 2020; Mikhaylov, 2020a, 2020b, 2020c; Yuan et al., 2021; Wang et al., 2019; Mikhaylov et al., 2021b; Mikhaylov, 2021a; Varyash et al., 2020; Zhao et al., 2021; An & Mikhaylov, 2020; Alwaelya et al., 2021; Yumashev & Mikhaylov, 2020; Yumashev et al., 2020; Mikhaylov et al., 2021a; Mutalimov et al., 2021; Morkovkin et al., 2020a, 2020b; An & Mikhaylov, 2021).

3 Results

Despite the fact that the energy risk insurance system is built on the basis of world experience and international practice and performs its main functions, this does not deprive it of a number of certain shortcomings that require the development and improvement of the system.

The fact that the energy risk insurance system has ceased to operate as steadily as it did before is evidenced by several facts at once: information on the official website of the Energy Risk Insurance Agency has ceased to be updated regularly, as it was before; information on the review of the energy risk market, on liquidation procedures and sanitized organizations has disappeared from public access (most often, recently, this information can only be found in certain periodicals).

All this contributes to the appearance of a cautious and anxious state of projects, which begin to suspect the Energy Risk Insurance Agency of providing false data and wanting to save on the interests of projects (Dinçer et al., 2019). If this trend continues, it is possible that the financial system will collapse and the energy risk insurance system will default in the event of a long-forgotten phenomenon such as "project panic." In order to avoid disrupting the stability of the energy risk insurance system, it is necessary to solve a number of problems, as well as to review certain points of federal legislation (Table 1).

It is still too early to talk about the complete overcoming of the crisis in the segment of term deposits of citizens. This is driven by two main nuances. Moreover, they also provoked the initial problems of these products.

Table 1 SWOT- analysis of the current system of energy risk insurance in Russia

Strengths	Weaknesses
• Formed insurance infrastructure. • Limited range of insurance protection objects. • Established relationships with credit institutions. • Consolidation and reorganization of the insurance business. • Experience of past crises. • High interest rates on deposits for companies. • Unwillingness of insurers to enter into such contracts.	• Low capitalization. • Low customer focus of the business. • Extremely high case management costs. • Low business profitability. • Low level of reliability. • Low level of staff qualification. • Unstable situation caused by the pandemic.
Opportunities	**Threats**
• Low level of insurance penetration. • Growing interest in the insurance industry on the part of the state. • Reform of the insurance industry supervision system. • Growth of investment attractiveness of individual insurers. • Potential benefit when applying for an insurance policy, since the payment of insurance may significantly exceed the amount of the deposit.	• Instability in global financial markets. • Insufficient quality of supervision of insurance companies. • Lack of supervision of insurance intermediaries. • Dumping. • Fraud.

First, a pandemic. Although it is on the wane, but everyone is preparing for the second wave. Moreover, its consequences may be even more negative. After all, not all business sectors and citizens recovered from the first one. Therefore, most likely, you will have to accumulate your savings again, directing them to everyday needs.

Secondly, the profitability of energy risks is steadily approaching the level of inflation, thus bringing the real profit of products to zero. That is, their use becomes relevant only for the safety of their savings, but not for multiplication. This pushes a number of Russians to extend their deposit agreements.

Leading Russian experts and scientists identify the following main problems related to the formation and functioning of the energy risk insurance system:

1. A fixed amount of state guarantees established at the state level (since the form of the Energy Risk Insurance Agency is state-owned), which is most often unfair in calculation and insufficient to meet the interests of projects. This implies non-compliance with the proportionality and direct dependence of the estimated amount of insurance compensation, in which, in the event of an increase in investment activity of the population, the amount of compensation remains unchanged, although logically it should increase, that is, it should have a dynamic character.
2. The system of contributions to the Mandatory Energy Risk Insurance Fund is outdated. All banks deduct contributions at a single basic percentage set by the Central Bank of the Russian Federation using the flat scale method, without taking into account the level of risks of each of them.

   For example, when considering the same Sberbank, it can be noted that the bank has a fairly high level of reliability, the risk of license revocation is minimized, and we can say that it is practically absent, since their energy risks are insured.

   But despite all this, it is this bank that is the main source of financing for the Mandatory Energy Risk Insurance Fund, since it forms a large part of it, by attracting a significant amount of deposits and deducting a percentage of insurance premiums calculated from the balances on deposit accounts. This is financially unprofitable for a bank, but it is undoubtedly very convenient for an Energy Risk Insurance Agency. This issue has already been raised several times by the management of several major Russian banks, including the Board of Sberbank, but so far it has remained unresolved.
3. The system of energy risk insurance in our country, unlike the world experience of using this system, covers a limited range of objects of insurance protection. Energy risk insurance systems in many foreign countries have been successfully insuring deposits of legal entities for quite a long time.
4. Insufficient public awareness of the operation of the energy risk insurance system and hidden information about the actual reliability of many banks. Despite the dissemination of information by the Energy Risk Insurance Agency, banks, and the mass media, there are still a considerable number of citizens in our country who do not know anything at all about the existence of the energy

risk insurance system, or have a minimum of information, and do not know how it functions until they face certain problems directly related to it.

In addition, some unscrupulous banks hide information about their unfavorable situation behind attractive rates, which is also deliberately not reported to the "ordinary" project. And many citizens simply do not trust banks, as they firmly believe that cash will be preserved better than in a deposit account, which indicates problems with financial literacy of the population as a whole.

5. Fraudulent actions in the field of energy risk insurance system. Among the main fraudulent schemes, there is the fragmentation of energy risks, as well as the fictitious formation of energy risks, with the aim of intentionally obtaining illegal insurance compensation.

   These schemes usually involve smaller banks that are part of the energy risk insurance system.

6. A very significant problem today is the policy of the Central Bank of the Russian Federation aimed at" improving" the banking sector. This measure, of course, is long overdue and necessary, but it should be borne in mind that the "recovery" procedure should be carried out more intelligently and expediently, without exposing the energy risk insurance system to risks.

   When banks are being rehabilitated and liquidated on a large scale, it is quite difficult to fill up the Mandatory Energy Risk Insurance Fund in a short time with only insurance premiums from participating banks.

   The Fund's insufficiency necessitates the use of borrowed funds by the Energy Risk Insurance Agency, which is always the beginning of the development of an even deeper and more difficult problem to solve. This is evidenced by the current statistics and forecast data determined for the future, which were discussed in more detail in the previous paragraph.

   Thus, it is worth noting that the current situation in the energy risk insurance market in our country requires urgent decision-making on many existing problems, improving the system itself and the legislative framework, as well as other possible options to minimize the risk of a system default, which may inevitably lead to the collapse of the country's banking system as a whole.

7. Fraud in the energy risk insurance system by bank employees. Bank managers and shareholders have learned to deceive supervisors and investors before they know it. The stolen funds are used by them to finance their personal businesses, as well as withdraw funds from projects to offshore zones.

   This is done through the use of fabricated reports, the creation of forged documents, and bribes offered to employees who acted as representatives of supervisory authorities, such as the Central Bank of the Russian Federation and the Energy Risk Insurance Agency.

8. The problem of off-balance sheet projects. When checking, the project data is not available in the bank's electronic databases, double-entry accounting is maintained, and, accordingly, the bank's balance sheet is distorted. As a result, when the bank's liquidation process begins, discrepancies between actual results and reported data are revealed, but this gap is already revealed too late.

9. Placement of deposits by large projects in foreign banks. The problem is that due to the limited amount of insurance compensation, energy risks in domestic banks are placed mainly by the middle class of citizens in terms of income. Large Russian millionaires and billionaires prefer foreign banks to Russian ones, namely those that guarantee 100% payment of insurance indemnities. This leads to an outflow of potential Russian deposits abroad, which also has an impact on the banking system as a whole.
10. Functioning of the energy risk insurance system in the Russian Federation without taking into account (with minimal consideration) moral hazard. Since the energy risk insurance system in our country is state-owned, and all responsibility for the occurrence of an insured event falls on the shoulders of the Energy Risk Insurance Agency, irresponsible behavior of other participants in the system is very often observed. On the part of banks, irresponsibility consists in insufficient informing customers about changes in their financial situation, as already mentioned earlier. Projects are irresponsible in choosing a bank to place a deposit with the maximum rates, without taking into account the degree of risk of the bank.

4 Conclusions and Discussion

In order to ensure the stable and uninterrupted operation of the energy risk insurance system, the Agency: pays compensation for deposits of individuals, including sole proprietors, as well as small businesses in the event of an insured event against a bank participating in the energy risk insurance system; maintains a register of banks participating in the energy risk insurance system; mandatory controls the formation of the energy risk Insurance Fund, including through bank contributions; manages the funds of the Energy Risk Insurance Fund. risks.

According to the data of the Energy Risk Insurance Agency, in 2019, the number of participants in energy risk insurance amounted to 723 banks. By 2020, this number has significantly decreased—to 696 (a decrease of 27 participants in the country compared to the previous period), including existing banks licensed to work with individuals in terms of attracting and placing deposits—372; existing credit institutions that previously accepted energy risks, but currently lost the right to attract funds from individuals –6; banks that have a license to work with individuals in terms of attracting and placing deposits -, in the process of liquidation—318.

In total, during the 15 years of operation of energy risk insurance, 481 insured events occurred, the total amount of insurance liability for which amounted to 1.92 trillion rubles for 9.3 million projects.

References

Alwaelya, S. A., Yousif, N. B. A., & Mikhaylov, A. (2021). Emotional development in preschoolers and socialization. *Early Child Development and Care, 190*(1717480).

An, J., & Mikhaylov, A. (2020). Russian energy projects in South Africa. *Journal of Energy in Southern Africa, 31*(3).

An, J., & Mikhaylov, A. (2021). Analysis of energy projects financial efficiency and renewable energy generation in Russia. *Finance: Theory and Practice, 25*(5), 1–11.

An, J., Mikhaylov, A., & Jung, S.-U. (2020b). The strategy of South Korea in the global oil market. *Energies, 13*(10), 2491.

An, J., Mikhaylov, A., & Jung, S.-U. (2021, March). A linear programming approach for robust network revenue management in the airline industry. *Journal of Air Transport Management, 91*, 101979.

An, J., Mikhaylov, A., & Kim, K. (2020c). Machine learning approach in heterogeneous group of algorithms for transport safety-critical system. *Applied Sciences, 10*(8), 2670.

An, J., Mikhaylov, A., & Moiseev, N. (2019b). Oil price predictors: Machine learning approach. *International Journal of Energy Economics and Policy, 9*(5), 1–6.

An, J., Mikhaylov, A., & Richter, U. H. (2020a). Trade war effects: Evidence from sectors of energy and resources in Africa. *Heliyon, 6*, e05693. https://doi.org/10.1016/j.heliyon.2020.e05693

An, J., Mikhaylov, A., & Sokolinskaya, N. (2019a). Machine learning in economic planning: Ensembles of algorithms. *Journal of Physics: Conference Series, 1353*, 012126.

Bhuiyan, M. A., An, J., Mikhaylov, A., Moiseev, N., & Danish, M. S. S. (2021). Renewable energy deployment and COVID-19 measures for sustainable development. *Sustainability, 13*, 4418. https://doi.org/10.3390/su13084418

Candila, V., Maximov, D., Mikhaylov, A., Moiseev, N., Senjyu, T., & Tryndina, N. (2021). On the relationship between oil and exchange rates of oil-exporting and oil-importing countries: From the great recession period to the Covid-19 era. *Energies, 14*, 8065.

Conteh, F., Takahashi, H., Hemeida, A. M., Krishnan, N., Mikhaylov, A., & Senjyu, T. (2021). Hybrid grid-connected renewable power generation for sustainable electricity supply in Sierra Leone: Case study lungi town. *Sustainability, 13*(18), 11435.

Danish, M. S. S., Bhattacharya, A., Stepanova, D., Mikhaylov, A., Grilli, M. L., Khosravy, M., & Senjyu, T. (2020). A systematic review of metal oxide applications for energy and environmental sustainability. *Metals, 10*(12), 1604.

Danish, M. S. S., Senjyu, T., Sabory, N. R., Khosravy, M., Grilli, M. L., Mikhaylov, A., & Majidi, H. (2021). A forefront framework for sustainable aquaponics modeling and design. *Sustainability, 13*, 9313. https://doi.org/10.3390/su13169313

Dayong, N., Mikhaylov, A., Bratanovsky, S., Shaikh, Z. A., & Stepanova, D. (2020). Mathematical modeling of the technological processes of catering products production. *Journal of Food Process Engineering, 43*(2), 1–9.

Denisova, V., Mikhaylov, A., & Lopatin, E. (2019). Blockchain infrastructure and growth of global power consumption. *International Journal of Energy Economics and Policy, 9*(4), 22–29.

Dinçer, H., Yüksel, S., Çağlayan, Ç., & Uluer, G. S. (2020). The contribution of nuclear energy investment on sustainable financial and economic development. *Journal of Financial Economics and Banking, 1*(1), 39–51.

Dinçer, H., Yüksel, S., & Martinez, L. (2019). Balanced scorecard-based analysis about European energy investment policies: A hybrid hesitant fuzzy decision-making approach with quality function deployment. *Expert Systems with Applications, 115*, 152–171.

Dong, B., Ikonnikova, I., Rogulin, R., Sakulyeva, T., & Mikhaylov, A. (2021). Environmental-economic approach to optimization of transport communication in megacities. *Journal of Environmental Science and Health, Part A, 56*(6), 660–666. https://doi.org/10.1080/10934529.2021.1913928

Dooyum, U. D., Mikhaylov, A., & Varyash, I. (2020). Energy security concept in Russia and South Korea. *International Journal of Energy Economics and Policy, 10*(4), 102–107.

Dorofeev, M. L. (2020). Impact of monetary policy on the level of economic inequality in the United States. *MGIMO Review of International Relations, 13*(5), 97–114. https://doi.org/10.24833/2071-8160-2020-5-74-97-114

Du, L., Dinçer, H., Ersin, İ., & Yüksel, S. (2020). IT2 fuzzy-based multidimensional evaluation of coal energy for sustainable economic development. *Energies, 13*(10), 2453.

Fang, S., Zhou, P., Dinçer, H., & Yüksel, S. (2021). Assessment of safety management system on energy investment risk using house of quality based on hybrid stochastic interval-valued intuitionistic fuzzy decision-making approach. *Safety Science, 141*, 105333.

Grilli, M. L., Valerini, D., Rizzo, A., Yilmaz, M., Song, C., Hu, G., Chierchia, R., Mikhaylov, A., & Rinaldi, A. (2021). A comparative study of the mechanical and tribological properties of thin Al_2O_3 coatings fabricated by atomic layer deposition and radiofrequency sputtering. *Physica Status Solidi (A), 218*(24), 1–25.

Gura, D., Mikhaylov, A., Glushkov, S., Zaikov, M., & Shaikh, Z. A. (2020). Model for estimating power dissipation along the interconnect length in single on-chip topology. *Evolutionary Intelligence*, s12065. https://doi.org/10.1007/s12065-020-00407-7

Huang, Y., Masrur, H., Shigenobu, R., Hemeida, A. M., Mikhaylov, A., & Senjyu, T. (2021a). A comparative Design of a Campus Microgrid Considering a multi-scenario and multi-objective approach. *Energies, 14*, 2853. https://doi.org/10.3390/en14112853

Huang, Y., Yona, A., Takahashi, H., Hemeida, A. M., Mandal, P., Mikhaylov, A., Senjyu, T., & Lotfy, M. E. (2021b). Energy management system optimization of drug store operate electric vehicles Charging Station. *Sustainability, 14*, 1217836.

Ivanyuk, V. (2018). Econometric forecasting models based on forecast combination methods. In *Proceedings of 2018 11th International Conference Management of Large-Scale System Development, MLSD 2018*. https://doi.org/10.1109/MLSD.2018.8551825

Ivanyuk, V., & Berzin, D. (2020). An algorithm for constructing an efficient investment portfolio. In R. Silhavy, P. Silhavy, & Z. Prokopova (Eds.), *Software engineering perspectives in intelligent systems. CoMeSySo 2020. Advances in intelligent systems and computing* (Vol. 1294). Springer. https://doi.org/10.1007/978-3-030-63322-6_39

Ivanyuk, V., & Levchenko, K. (2020). Intelligent methods for predicting financial time series. In R. Silhavy, P. Silhavy, & Z. Prokopova (Eds.), *Software engineering perspectives in intelligent systems. CoMeSySo 2020. Advances in intelligent systems and computing* (Vol. 1294). Springer. https://doi.org/10.1007/978-3-030-63322-6_41

Ivanyuk, V., & Soloviev, V. (2019). Efficiency of neural networks in forecasting problems. In *Proceedings of 2019 12th International Conference Management of Large-Scale System Development; MLSD 2019*. https://doi.org/10.1109/MLSD.2019.8911046

Ivanyuk, V., Sunchalin, A., & Sunchalina, A. (2020). Development of an intelligent ensemble forecasting system. In R. Silhavy, P. Silhavy, & Z. Prokopova (Eds.), *Software engineering perspectives in intelligent systems. CoMeSySo 2020. Advances in intelligent systems and computing* (Vol. 1294). Springer. https://doi.org/10.1007/978-3-030-63322-6_40

Li, X., Zhu, S., Yüksel, S., Dinçer, H., & Ubay, G. G. (2020). Kano-based mapping of innovation strategies for renewable energy alternatives using hybrid interval type-2 fuzzy decision-making approach. *Energy, 211*, 118679.

Lisin, A. (2020). Valuation of the activities of foreign banks in the Russian banking sector. *Orbis, 15*(45), 53–63. http://www.revistaorbis.org.ve/pdf/45/art5.pdf

Liu, Z.-J., Panfilova, E., Mikhaylov, A., & Kurilova, A. (2021). COVID-19 crisis impact on the stability between parties in crowdfunding and crowdsourcing. *Wireless Personal Communications, 8932*. https://doi.org/10.1007/s11277-021-08932-z

Liu, Z.-J., Panfilova, E., Mikhaylov, A., & Kurilova, A. (2022). Assessing stability in relationship between parties in crowdfunding and crowdsourcing projects during the COVID-19 crisis. *Journal of Global Information Management, 30*(2), 1.

Melnichuk, A. V., Melnichuk, Y. A., Bocharov, A. V., Stepanova, D. I., & Kurbanov, S. A. (2020). Constitutional foundations of the functioning of local self-government. *Revista Inclusiones, 7*, 337–346.

Meynkhard, A. (2019). Fair market value of bitcoin: Halving effect. *Investment Management and Financial Innovations, 16*(4), 72–85. https://doi.org/10.21511/imfi.16(4).2019.07

Meynkhard, A. (2020). Effect of bitcoin volatility on altcoins pricing. *Advances in Intelligent Systems and Computing, 1294*, 652–664. https://doi.org/10.1007/978-3-030-63322-6_55

Mikhaylov, A. (2018a). Volatility spillover effect between stock and exchange rate in oil exporting countries. *International Journal of Energy Economics and Policy, 8*(3), 321–326.

Mikhaylov, A. (2018b). Pricing in oil market and using probit model for analysis of stock market effects. *International Journal of Energy Economics and Policy, 8*(2), 69–73.

Mikhaylov, A. (2018c). The yield of the carry trade strategy. *Finance: Theory and Practice, 22*(3), 52–63. https://doi.org/10.26794/2587-5671-2018-22-3-52-63

Mikhaylov, A. (2020a). Cryptocurrency market analysis from the open innovation perspective. *Journal of Open Innovation: Technology, Market, and Complexity, 6*(4), 197. https://doi.org/10.3390/joitmc6040197

Mikhaylov, A. (2020b). Geothermal energy development in Iceland. *International Journal of Energy Economics and Policy, 10*(4), 31–35.

Mikhaylov, A. (2020c). Cryptocurrency market development: Hurst method. *Finance: Theory and Practice, 24*(3), 81–91.

Mikhaylov, A. (2021a). Development of Friedrich von Hayek's theory of private money and economic implications for digital currencies. *Terra Economicus, 19*(1), 53–62.

Mikhaylov, A. (2021b). Lichens as indicators of atmospheric pollution in urban ecosystems. *Israel Journal of Ecology & Evolution, 67*(1–2), 60–68. https://doi.org/10.1163/22244662-bja10016

Mikhaylov, A. (2022). Efficiency of renewable energy generation in Russia. *Annals of the Brazilian Academy of Sciences, 94*(3).

Mikhaylov, A., Burakov, D., & Didenko, V. (2019). Relationship between oil Price and macroeconomic indicators in Russia. *Finance: Theory and Practice, 23*(2), 105–116. https://doi.org/10.26794/2587-5671-2019-23-2-105-116

Mikhaylov, A., Danish, M. S. S., & Senjyu, T. (2021b). New stage in evolution of cryptocurrency market: Analysis by Hurst method. In H. Dincer & S. Yuksel (Eds.), *Strategic outlook in business and finance innovation: Multidimensional policies for emerging economies* (pp. 35–45. ISBN: 978-1-80043-445-5). Emerald Publishing Limited. https://doi.org/10.1108/978-1-80043-444-820211004

Mikhaylov, A., Moiseev, N., Aleshin, K., & Burkhardt, T. (2020). Global climate change and greenhouse effect. *Entrepreneurship and Sustainability Issues, 7*(4), 2897–2913.

Mikhaylov, A., Sokolinskaya, N., & Nyangarika, A. (2018). Optimal carry trade strategy based on currencies of energy and developed economies. *Journal of Reviews on Global Economics, 7*, 582–592.

Mikhaylov, A., & Tarakanov, S. (2020). Development of Levenberg-Marquardt theoretical approach for electric network. *Journal of Physics: Conference Series, 1515*, 052006.

Mikhaylov, A., Yumashev, A., & Kolpak, E. (2021a). Quality of life, anxiety and depressive disorders in patients with extrasystolic arrhythmia. *Archives of Medical Science, 17*(1). https://doi.org/10.5114/aoms.2020.101359

Moiseev, N., Mikhaylov, A., Varyash, I., & Saqib, A. (2020). Investigating the relation of GDP per capita and corruption index. *Entrepreneurship and Sustainability Issues, 8*(1), 780–794.

Moiseev, N., Sorokin, A., Zvezdina, N., Mikhaylov, A., Khomyakova, L., & Danish, M. S. S. (2021). Credit risk theoretical model on the base of DCC-GARCH in time-varying parameters framework. *Mathematics, 9*, 2423. https://doi.org/10.3390/math9192423

Morkovkin, D. E., Gibadullin, A. A., Kolosova, E. V., Semkina, N. S., & Fasehzoda, I. S. (2020a). Modern transformation of the production base in the conditions of industry 4.0: Problems and prospects. *Journal of Physics: Conference Series, 1515*, 032014.

Morkovkin, D., Lopatkin, D., Sadriddinov, M., Shushunova, T., Gibadullin, A., & Golikova, O. (2020b). Assessment of innovation activity in the countries of the world. *E3S Web of Conferences, 157*, 04015. https://doi.org/10.1051/e3sconf/202015704015

Mukhametov, A., Bekhorashvili, A., Avdeenko, A., & Mikhaylov, A. (2021). The impact of growing legume plants under conditions of Biologization and soil cultivation on Chernozem fertility and productivity of rotation crops. *Legume Research, 44*, 12.

Mutalimov, V., Kovaleva, I., Mikhaylov, A., & Stepanova, D. (2021). Assessing regional growth of small business in Russia. *Entrepreneurial Business and Economics Review, 9*(3), 119–133. https://doi.org/10.15678/EBER.2021.090308

Nie, D., Panfilova, E., Samusenkov, V., & Mikhaylov, A. (2020). E-learning financing models in Russia for sustainable development. *Sustainability, 12*(11), 4412.

Nyangarika, A., Mikhaylov, A., & Richter, U. (2019a). Influence oil price towards economic indicators in Russia. *International Journal of Energy Economics and Policy, 9*(1), 123–129.

Nyangarika, A., Mikhaylov, A., & Richter, U. (2019b). Oil price factors: Forecasting on the base of modified auto-regressive integrated moving average model. *International Journal of Energy Economics and Policy, 9*(1), 149–159.

Nyangarika, A., Mikhaylov, A., & Tang, B.-J. (2018). Correlation of oil prices and gross domestic product in oil producing countries. *International Journal of Energy Economics and Policy, 8*(5), 42–48.

Qiu, D., Dinçer, H., Yüksel, S., & Ubay, G. G. (2020). Multi-faceted analysis of systematic risk-based wind energy investment decisions in E7 economies using modified hybrid modeling with IT2 fuzzy sets. *Energies, 13*(6), 1423.

Radosteva, M., Soloviev, V., Ivanyuk, V., & Tsvirkun, A. (2018). Use of neural network models in market risk management. *Advances in Systems Science and Applications, 18*(2), 53–58. https://doi.org/10.25728/assa.2018.18.2.582

Ranjbar, O., Chang, T., Nel, E., & Gupta, R. (2017). Energy consumption and economic growth nexus in South Africa: Asymmetric frequency domain approach. *Energy Sources, Part B: Economics, Planning, and Policy, 12*(1), 24–31.

Rathnayaka, R. K. T., Seneviratna, D. M. K., & Long, W. (2018). The dynamic relationship between energy consumption and economic growth in China. *Energy Sources, Part B: Economics, Planning, and Policy, 13*(5), 264–268.

Saqib, A., Chan, T.-H., Mikhaylov, A., & Lean, H. H. (2021). Are the responses of sectoral energy imports asymmetric to exchange rate volatilities in Pakistan? Evidence from recent foreign exchange regime. *Frontiers in Energy Research, 9*, 614463. https://doi.org/10.3389/fenrg.2021.614463

Sunchalin, A. M., Kochkarov, R. A., Levchenko, K. G., Kochkarov, A. A., & Ivanyuk, V. A. (2019). Methods of risk management in portfolio theory. *Espacios, 40*(16), 8.

Uandykova, M., Lisin, A., Stepanova, D., Baitenova, L., Mutaliyeva, L., Yuksel, S., & Dincer, H. (2020). The social and legislative principles of counteracting ransomware crime. *Entrepreneurship and Sustainability Issues, 8*(2), 777–798.

Udalov, I. (2021). The transition to renewable energy sources as a threat to resource economies. *International Journal of Energy Economics and Policy, 11*(3), 1–6.

Uyeh, D. D., Asem-Hiablie, S., Park, T., Kim, K.-M., Mikhaylov, A., Woo, S., & Ha, Y. (2021). Could japonica Rice be an alternative variety for increased global food security and climate change mitigation? *Food, 10*, 1869. https://doi.org/10.3390/foods10081869

Varyash, I., Mikhaylov, A., Moiseev, N., & Aleshin, K. (2020). Triple bottom line and corporate social responsibility performance indicators for Russian companies. *Entrepreneurship and Sustainability Issues, 8*(1), 313–331.

Wang, S., Liu, Q., Yuksel, S., & Dincer, H. (2019). Hesitant linguistic term sets-based hybrid analysis for renewable energy investments. *IEEE Access, 7*, 114223–114235.

Yuan, G., Xie, F., Dinçer, H., & Yüksel, S. (2021). The theory of inventive problem solving (TRIZ)-based strategic mapping of green nuclear energy investments with spherical fuzzy group decision-making approach. *International Journal of Energy Research., 45*(8), 12284–12300.

Yüksel, S., Dinçer, H., Çağlayan, Ç., & Uluer, G. S. (2021a). Determining the optimal financial strategies for nuclear energy companies. In *Financial strategies in competitive markets* (pp. 1–16). Springer.

Yüksel, S., Khomyakova, L., & Mikhaylov, A. (2021b). Energy center selection in G7 industry with fuzzy MOORA. In S. Yüksel & H. Dinçer (Eds.), *Handbook of research on strategic Management for Current Energy Investments* (pp. 87–106). IGI Global. isbn 13: 9781799883357. https://doi.org/10.4018/978-1-7998-8335-7.ch006

Yüksel, S., Mikhaylov, A., & Ubay, G. G. (2021d). Factors causing delay in the installation of nuclear power plants. In S. Yüksel & H. Dinçer (Eds.), *Strategic approaches to energy management. Contributions to management science*. Springer. https://doi.org/10.1007/978-3-030-76783-9_7

Yüksel, S., Mikhaylov, A., Ubay, G. G., & Uyeh, D. D. (2021c). The role of hydrogen in the Black Sea for the future energy supply security of Turkey. In S. Yüksel & H. Dinçer (Eds.), *Handbook of research on strategic Management for Current Energy Investments* (pp. 1–15). IGI Global. https://doi.org/10.4018/978-1-7998-8335-7.ch001

Yumashev, A., & Mikhaylov, A. (2020). Development of polymer film coatings with high adhesion to steel alloys and high wear resistance. *Polymer Composites, 41*(7), 2875–2880.

Yumashev, A., Ślusarczyk, B., Kondrashev, S., & Mikhaylov, A. (2020). Global indicators of sustainable development: Evaluation of the influence of the human development index on consumption and quality of energy. *Energies, 13*, 2768.

Zhao, Q., Cherkasov, S., Avdeenko, A., Kondratenko, L., & Mikhaylov, A. (2021). Integral estimate of the added value of agricultural crops in the synergetic agriculture on the example of vicia faba (Vicia faba L.). *Polish Journal of Environmental Studies, 30*(5), 1–9. https://doi.org/10.15244/pjoes/131949

Zhou, P., Luo, J., Cheng, F., Yüksel, S., & Dinçer, H. (2021). Analysis of risk priorities for renewable energy investment projects using a hybrid IT2 hesitant fuzzy decision-making approach with alpha cuts. *Energy, 224*, 120184.

How to Improve Energy Investments in Russia

Elizaveta Ibragimova and Mir Sayed Shah Danish

1 Introduction

The extension of insurance coverage to legal entities will be a stimulating factor for such lenders to accumulate financial resources in the banking sector of the Russian Federation. Expanding the scope of insurance will require a review of the current model of the energy risk insurance system, which currently includes only those banks whose license provides for the right to insure energy risks. Extending insurance coverage to legal entities, while maintaining the current conditions for mandatory participation in the energy risk insurance system, on the one hand, may contribute to the outflow of funds of legal entities from credit institutions that are not participants in the energy risk insurance system, and on the other hand, it may be possible to increase the number of employees who are not participants in the—may lead to the formation of a risky business model for such credit institutions.

This creates a moral hazard that depends on the effectiveness of market discipline. It is assumed that an effective market discipline that reduces the degree of moral hazard is a situation in which the project carefully monitors the financial condition of the bank, and, if there is a risk of loss of funds, begins to demand an increase in the deposit rate, or withdraws money from this bank altogether. Thus, there is a change in the structure of the bank's financial assets, which reduces its stability. The main thing in this mechanism is the close attention of the project to the bank. Now, since the project already knows that its money is protected, this mechanism does not work,

E. Ibragimova (✉)
Financial Research Institute of the Ministry of Finance of the Russian Federation, Moscow, Russia

M. S. S. Danish
Strategic Research Projects Center, University of the Ryukyus, Okinawa, Japan

© The Author(s), under exclusive license to Springer Nature Switzerland AG 2022

H. Dinçer, S. Yüksel (eds.), *Sustainability in Energy Business and Finance*, Contributions to Finance and Accounting,
https://doi.org/10.1007/978-3-030-94051-5_12

which leads to certain problems and imposes additional burden and responsibility on the control bodies.

It is worth noting that what was in operation 15–20 years ago is no longer suitable for functioning at the present time. That is why it is necessary to make drastic changes in the processes of formation and functioning of the system, which must necessarily have legislative consolidation.

So, there is no need to" reinvent the wheel," since there are many types and interpretations of energy risk insurance systems that are successfully used by other countries, it is only necessary to shift this experience to Russian conditions. It is these urgent changes that will be discussed in the next chapter of this study.

2 Literature Review

Taking into account international experience, it is advisable to expand the range of persons (entities) covered by the protection provided by the energy risk insurance system by including legal entities' accounts and deposits in the system (Bhuiyan et al., 2021; Dong et al., 2021, Mikhaylov, 2021b; Liu et al., 2022; Saqib et al., 2021; Radosteva et al., 2018; Ranjbar et al., 2017; Rathnayaka et al., 2018; Sunchalin et al., 2019; Uandykova et al., 2020; Udalov, 2021; Yüksel et al., 2021a, 2021b, 2021c; 2021d; Dinçer et al., 2019; Dorofeev, 2020).

We believe that funds of legal entities whose activities are related to the provision of financial services and are based, in particular, on a professional assessment of the stability of financial institutions and counterparties should be excluded from the scope of energy risk insurance (Mikhaylov, 2018c; Dinçer et al., 2020; Qiu et al., 2020; Mikhaylov et al., 2019; Melnichuk et al., 2020; Nie et al., 2020; Moiseev et al., 2021; Zhou et al., 2021; Grilli et al., 2021).

In this regard, it is proposed not to extend insurance coverage to bank accounts (deposits) of credit institutions; professional participants in the securities market; trade organizers; clearing organizations; microfinance organizations; consumer credit cooperatives; insurance organizations; insurance brokers; mutual insurance companies; non-state pension funds; management companies of investment funds (mutual investment funds) and non-state pension funds; specialized depositories of investment funds (mutual investment funds). private pension funds; pawnshops; leasing companies (Dayong et al. 2020; Mikhaylov et al., 2018; Nyangarika et al., 2018; Danish et al., 2020, 2021; Fang et al., 2021; Li et al., 2020; Lisin, 2020; An et al., 2021; Ivanyuk & Berzin, 2020; Ivanyuk & Levchenko, 2020; Ivanyuk et al., 2020; Ivanyuk, 2018; Ivanyuk & Soloviev, 2019; Du et al., 2020; Uyeh et al., 2021).

Taking into account international experience and recommendations, it is also not advisable to insure funds of the federal budget placed in credit institutions, funds of budgets of constituent entities of the Russian Federation and local budgets, funds of state and other extra-budgetary funds, with the exception of funds of state and municipal social institutions (for example, educational and medical institutions). It is debatable to include "funds in settlements" (transfers without opening an account, letters of credit, etc.) in the perimeter of the energy risk insurance system, taking into account (Yuan et al., 2021; Wang et al., 2019): accumulation of such funds on bank accounts for a relatively short period of time. The inclusion of bank accounts (energy risks) of individuals in precious metals in the insurance coverage period seems impractical due to the predominantly investment nature of such accounts and, as a consequence, their use by investors who are able to assess certain risks associated with the dependence of profitability on such accounts (deposits) on market quotations for metal placed on accounts (deposits) (Denisova et al., 2019; Nyangarika et al., 2019a, 2019b; Huang et al., 2021a, 2021b; Mikhaylov, 2018a, 2018b, 2022; Meynkhard, 2019; Mikhaylov et al., 2019; Conteh et al., 2021; Meynkhard, 2020).

In addition, it does not seem appropriate to include individual legal entities in the insurance perimeter of accounts (deposits), for which the limit of insurance compensation is generally insignificant (An et al., 2019a, 2019b, 2020a, 2020b, 2020c; Mikhaylov, 2019, 2020a, 2020b, 2020c, 2021a; Mikhaylov & Tarakanov, 2020; Moiseev et al., 2020; Gura et al., 2020; Dooyum et al., 2020; Mikhaylov et al., 2021b; Varyash et al., 2020; Zhao et al., 2021; An & Mikhaylov, 2020; Alwaelya et al., 2021; Yumashev & Mikhaylov, 2020; Yumashev et al., 2020; Mikhaylov et al., 2021b; Mutalimov et al., 2021; Morkovkin et al., 2020a, 2020b; An & Mikhaylov, 2021; Mukhametov et al., 2021; Candila et al., 2021; Liu et al., 2021).

In this regard, two possible solutions can be considered for including legal entities 'accounts in the insurance perimeter: the first option: including only the accounts of medium-sized enterprises (in addition to small ones), state and municipal unitary enterprises, state and municipal budget institutions of a social nature (for example, educational and medical institutions) in the insurance perimeter; the second option: including all legal entities in the insurance perimeter, except for legal entities that are executors (lead executors) of the state defense order that have opened there are separate accounts with authorized banks, and business strategic companies, with the exception of those related to small and medium-sized enterprises, as well as state-owned corporations.

In order to gradually adapt the banking sector to major regulatory changes, it is possible to consider the feasibility of gradually reforming the energy risk insurance system.

3 Methods

At the first stage, insurance coverage can be extended to funds placed in credit institutions by certain types of socially oriented non-commercial organizations and non-profit associations of citizens: non-profit organizations operating in one of the following organizational and legal forms (horticultural non-profit partnerships; horticultural non-profit partnerships; homeowners 'associations; garage and garage-building cooperatives; housing and housing-building cooperatives; Cossack societies included in the state register of Cossack associations). Russian Federation and registered as a legal entity; communities of small indigenous peoples of the

Russian Federation registered as a legal entity; religious organizations) non-profit organizations performing socially useful services that meet the requirements established by Federal Law No. 7-FZ of 12.01.1996 "On Non—Profit Organizations," information about which is contained in the register of non-profit organizations performing socially useful services, maintained in accordance with the specified Federal Law.

In addition, at the first stage, we believe it is possible to include in the insurance perimeter special accounts opened in accordance with the norms of the Housing Code of the Russian Federation, where funds of the capital repair fund are placed (in fact, these are contributions of individuals for capital repairs of an apartment building), regardless of who is the owner of the special account of the capital repair fund.

At the first stage, it is also advisable to establish an increased limit of insurance compensation in the amount of ten million rubles, paid to individual projects in the following special life situations: sale of residential premises and (or) land plot (part of land plot) on which a residential building (part of a residential building), garden house (part of a garden house) is located; receipt of inheritance; compensation for damage caused to life, health, or personal property, receipt of social payments and benefits; execution of a court decision; receipt of grants in the form of subsidies; receipt of funds from charitable organizations by collecting donations or other voluntary targeted transfers for the treatment of serious illnesses of the project or its close relatives; expenses related to a serious illness of the project or a member of its family. At the same time, the individual project's right to receive an increased amount of insurance compensation for the specified reasons will be retained only for three months from the date of crediting the corresponding funds to the project account.

According to preliminary estimates of the state corporation "Energy Risk Insurance Agency," the transformation of the mechanism for forming the mandatory energy risk insurance fund in connection with these changes will not be required.

At the next stage, the Bank of Russia is considering the possibility of including other legal entities in the perimeter of the energy risk insurance system and, taking into account the need for consistent regulation, the possibility of including funds deposited in the accounts of notaries, lawyers, and other persons whose accounts are open for professional activities.

4 Results

We can distinguish the following proposals for the development of the Russian energy risk insurance system:

- Voluntary deposit insurance of the corporate sector may become popular if insurers develop favorable conditions and adequate tariffs for these services.

- Further development of such a function of the energy agency as financial rehabilitation of commercial banks will help to improve the situation in problem banks even before it becomes irreversible and entails the payment of project reimbursements from the fund.
- Voluntary insurance of energy risks of individuals (the amount of excess over the amount subject to mandatory energy risk insurance) will allow you to keep temporarily available funds in one bank.
- Improving the financial education of citizens through training, informing the general public in the field of banking services for individuals, in particular regarding energy risk insurance.
- Improving the legal system of energy risk insurance and drawing up an action plan in case of a crisis of the entire banking system implies further development of the policy and legislative framework for a mandatory and structured system of energy risk insurance.

In general, the established energy risk insurance regime is characterized as a positive innovation that ensures the stability and sustainability of the Russian banking system. The trust of projects that now have the state-backed confidence that in the event of a credit institution's bankruptcy, they will receive a refund of the invested funds is restored. The inflow of energy risks to private banks is increasing, and the scale of financial intermediation of the country's banking system is increasing.

This also has additional advantages for credit institutions: the social responsibility of banks, which consists in ensuring the safety of citizens' savings, is being strengthened. Among other things, the system is a kind of mechanism for ensuring its own security and reducing the level of risks.

Thus, the insurance of bank energy risks has the following trends today:

1. Remote communication channels with customers and participants of energy risk insurance are actively developing.
2. The management system of the energy agency is being improved in terms of ensuring transparency and openness of the Agency's activities in order to attract the largest number of projects and strengthen the trust of existing ones.
3. There is an expansion of the functional responsibilities of the Energy Risk Insurance Agency in terms of the full transfer of powers of the role of sanator and liquidator of banks from the Bank of Russia to the Energy Agency.
4. The Agency implements its own standards in terms of ensuring the sustainable functioning of the banking energy risk insurance system in the country and abroad.
5. It is planned to expand the list of objects of insurance protection by including legal entities in it.
6. Introduction of new and expansion of existing information portals (mass media, newspapers, advertisements, reference books, leaflets, etc.) dedicated to the existence of a system of insurance of bank energy risks (deposits) and the mechanism of functioning of this system.

7. Holding exhibitions, fairs, seminars, presentations in terms of highlighting new products in the field of energy risk insurance, as well as dissemination of information about the possibility of participating in this system.
8. Expansion and establishment of partnership relations of the Agency outside the country with international organizations (Federal Deposit Insurance Corporation in the USA, State Energy Risk Insurance Fund in Germany, Energy Risk Insurance Fund in Turkey, Energy Risk Guarantee Fund in Poland, etc.).
9. Creation of a system of emergency payments related to the increase in the volume of insurance claims during the crisis, economic stagnation, etc.
10. Minimize and assess possible risks associated with the financial stability of the energy risk insurance system in order to increase liquidity.

In conclusion, I would like to say that the deposit insurance system is very important for the banking system and for the economy as a whole. Because this system acts as a guarantee of the security of monetary savings of the country's population.

For the sustainability of this system, it is recommended to conduct a competent and phased monetary policy of the state in terms of insurance, as well as to increase the financial literacy of the population by creating a transparent insurance system in general.

Thus, these measures will increase not only the economic well-being of the country, but also the level of public confidence, which is very important for the successful development of the country.

5 Conclusions and Discussion

Despite the fact that the energy risk insurance system is built on the basis of world experience and international practice and performs its main functions, this does not deprive it of a number of certain shortcomings that require the development and improvement of the system. In order to avoid disrupting the stability of the energy risk insurance system, it is necessary to solve a number of problems, as well as to review certain points of federal legislation.

It is worth noting that what was in operation 15–20 years ago is no longer suitable for functioning at the present time. That is why it is necessary to make drastic changes in the processes of formation and functioning of the system, which must necessarily have legislative consolidation.

Taking into account international experience, it is advisable to expand the range of persons (entities) covered by the protection provided by the energy risk insurance system by including legal entities' accounts and deposits in the system. The extension of insurance coverage to legal entities will be a stimulating factor for such lenders to accumulate financial resources in the banking sector of the Russian Federation.

Expanding the scope of insurance will require a review of the current model of the energy risk insurance system, which currently includes only those banks whose license provides for the right to insure energy risks.

We believe that funds of legal entities whose activities are related to the provision of financial services and are based, in particular, on a professional assessment of the stability of financial institutions and counterparties should be excluded from the scope of energy risk insurance.

In this regard, it is proposed not to extend insurance coverage to bank accounts (deposits) of credit institutions; professional participants in the securities market; trade organizers; clearing organizations; microfinance organizations; consumer credit cooperatives; insurance organizations; insurance brokers; mutual insurance companies; non-state pension funds; management companies of investment funds (mutual investment funds) and non-state pension funds; specialized depositories of investment funds (mutual investment funds). private pension funds; pawnshops; leasing companies.

The following proposals can be identified for the development of the Russian energy risk insurance system: voluntary deposit insurance for the corporate sector may become in demand if insurers develop favorable conditions and adequate tariffs for these services; further development of the energy agency's function as financial rehabilitation of commercial banks will help improve the situation in problem banks even before it becomes irreversible and entails the payment of project reimbursements from the fund; voluntary energy agency insurance increasing the financial education of citizens through training, informing the general public in the field of banking services for individuals, in particular regarding energy risk insurance; improving the legal system of energy risk insurance and drawing up an action plan in case of a crisis of the entire banking system implies further development of the political and economic policy of the entire banking system. and the legal framework for a mandatory and structured system of energy risk insurance. Insufficient supervision of banks participating in the system, as well as new banks allowed to participate in the system. At first glance, the selection criteria for banks to participate in the energy risk insurance system are quite strict, and the verification of banks is quite thorough. But, judging by the results of statistics, after some time, banks that were initially recognized as financially stable and successfully developing in order to participate in the system are recognized as insolvent. The reason why this is happening is obvious, and the supervision was not strict enough. This problem is undoubtedly serious and needs to be solved, since without this, there are many banks that are undergoing the process of rehabilitation or liquidation. The presence of such an additional problem leads not so much to the replenishment of the Mandatory Energy Risk Insurance Fund by new participating banks but to the cost of paying insurance indemnities. Naturally, all the problems listed above are far from the only ones, but they belong to the main and, in principle, generalizing ones. Based on the information presented in this chapter, it can be concluded that the insurance system that operates in our country today, almost in its original form, is already beginning to lose its relevance, as rapid changes in the insurance market are taking place, what is happening in the economy requires the same rapid response in the form of improving the energy risk insurance system and adapting it to new conditions.

References

Alwaelya, S. A., Yousif, N. B. A., & Mikhaylov, A. (2021). Emotional development in preschoolers and socialization. *Early Child Development and Care, 190*(1717480).
An, J., & Mikhaylov, A. (2020). Russian energy projects in South Africa. *Journal of Energy in Southern Africa, 31*(3), 58–64.
An, J., & Mikhaylov, A. (2021). Analysis of energy projects financial efficiency and renewable energy generation in Russia. *Finance: Theory and Practice, 25*(5), 1–11.
An, J., Mikhaylov, A., & Jung, S.-U. (2020b). The strategy of South Korea in the global oil market. *Energies, 13*(10), 2491.
An, J., Mikhaylov, A., & Jung, S.-U. (2021, March). A linear programming approach for robust network revenue management in the airline industry. *Journal of Air Transport Management, 91*, 101979.
An, J., Mikhaylov, A., & Kim, K. (2020c). Machine learning approach in heterogeneous group of algorithms for transport safety-critical system. *Applied Sciences, 10*(8), 2670.
An, J., Mikhaylov, A., & Moiseev, N. (2019b). Oil price predictors: Machine learning approach. *International Journal of Energy Economics and Policy, 9*(5), 1–6.
An, J., Mikhaylov, A., & Richter, U. H. (2020a). Trade war effects: Evidence from sectors of energy and resources in Africa. *Heliyon, 6*, e05693. https://doi.org/10.1016/j.heliyon.2020.e05693
An, J., Mikhaylov, A., & Sokolinskaya, N. (2019a). Machine learning in economic planning: Ensembles of algorithms. *Journal of Physics: Conference Series, 1353*, 012126.
Bhuiyan, M. A., An, J., Mikhaylov, A., Moiseev, N., & Danish, M. S. S. (2021). Renewable energy deployment and COVID-19 measures for sustainable development. *Sustainability, 13*, 4418. https://doi.org/10.3390/su13084418
Candila, V., Maximov, D., Mikhaylov, A., Moiseev, N., Senjyu, T., & Tryndina, N. (2021). On the relationship between oil and exchange rates of oil-exporting and oil-importing countries: From the great recession period to the Covid-19 era. *Energies, 14*, 8065.
Conteh, F., Takahashi, H., Hemeida, A. M., Krishnan, N., Mikhaylov, A., & Senjyu, T. (2021). Hybrid grid-connected renewable power generation for sustainable electricity supply in Sierra Leone: Case study lungi town. *Sustainability, 13*(18), 11435.
Danish, M. S. S., Bhattacharya, A., Stepanova, D., Mikhaylov, A., Grilli, M. L., Khosravy, M., & Senjyu, T. (2020). A systematic review of metal oxide applications for energy and environmental sustainability. *Metals, 10*(12), 1604.
Danish, M. S. S., Senjyu, T., Sabory, N. R., Khosravy, M., Grilli, M. L., Mikhaylov, A., & Majidi, H. (2021). A forefront framework for sustainable aquaponics modeling and design. *Sustainability, 13*, 9313. https://doi.org/10.3390/su13169313
Dayong, N., Mikhaylov, A., Bratanovsky, S., Shaikh, Z. A., & Stepanova, D. (2020). Mathematical modeling of the technological processes of catering products production. *Journal of Food Process Engineering, 43*(2), 1–9.
Denisova, V., Mikhaylov, A., & Lopatin, E. (2019). Blockchain infrastructure and growth of global power consumption. *International Journal of Energy Economics and Policy, 9*(4), 22–29.
Dinçer, H., Yüksel, S., Çağlayan, Ç., & Uluer, G. S. (2020). The contribution of nuclear energy investment on sustainable financial and economic development. *Journal of Financial Economics and Banking, 1*(1), 39–51.
Dinçer, H., Yüksel, S., & Martinez, L. (2019). Balanced scorecard-based analysis about European energy investment policies: A hybrid hesitant fuzzy decision-making approach with quality function deployment. *Expert Systems with Applications, 115*, 152–171.

Dong, B., Ikonnikova, I., Rogulin, R., Sakulyeva, T., & Mikhaylov, A. (2021). Environmental-economic approach to optimization of transport communication in megacities. *Journal of Environmental Science and Health, Part A, 56*(6), 660–666. https://doi.org/10.1080/10934529.2021.1913928

Dooyum, U. D., Mikhaylov, A., & Varyash, I. (2020). Energy security concept in Russia and South Korea. *International Journal of Energy Economics and Policy, 10*(4), 102–107.

Dorofeev, M. L. (2020). Impact of monetary policy on the level of economic inequality in the United States. *MGIMO Review of International Relations, 13*(5), 97–114. https://doi.org/10.24833/2071-8160-2020-5-74-97-114

Du, L., Dinçer, H., Ersin, İ., & Yüksel, S. (2020). IT2 fuzzy-based multidimensional evaluation of coal energy for sustainable economic development. *Energies, 13*(10), 2453.

Fang, S., Zhou, P., Dinçer, H., & Yüksel, S. (2021). Assessment of safety management system on energy investment risk using house of quality based on hybrid stochastic interval-valued intuitionistic fuzzy decision-making approach. *Safety Science, 141*, 105333.

Grilli, M. L., Valerini, D., Rizzo, A., Yilmaz, M., Song, C., Hu, G., Chierchia, R., Mikhaylov, A., & Rinaldi, A. (2021). A comparative study of the mechanical and tribological properties of thin Al2O3 coatings fabricated by atomic layer deposition and radiofrequency sputtering. *Physica Status Solidi (A), 218*(24), 1–25.

Gura, D., Mikhaylov, A., Glushkov, S., Zaikov, M., & Shaikh, Z. A. (2020). Model for estimating power dissipation along the interconnect length in single on-chip topology. *Evolutionary Intelligence*, s12065.

Huang, Y., Masrur, H., Shigenobu, R., Hemeida, A. M., Mikhaylov, A., & Senjyu, T. (2021a). A comparative Design of a Campus Microgrid Considering a multi-scenario and multi-objective approach. *Energies, 14*, 2853. https://doi.org/10.3390/en14112853

Huang, Y., Yona, A., Takahashi, H., Hemeida, A. M., Mandal, P., Mikhaylov, A., Senjyu, T., & Lotfy, M. E. (2021b). Energy management system optimization of drug store operate electric vehicles Charging Station. *Sustainability, 14*, 1217836.

Ivanyuk, V. (2018). Econometric forecasting models based on forecast combination methods. In *Proceedings of 2018 11th International Conference Management of Large-Scale System Development, MLSD 2018*. https://doi.org/10.1109/MLSD.2018.8551825

Ivanyuk, V., & Berzin, D. (2020). An algorithm for constructing an efficient investment portfolio. In R. Silhavy, P. Silhavy, & Z. Prokopova (Eds.), *Software engineering perspectives in intelligent systems. CoMeSySo 2020. Advances in intelligent systems and computing* (Vol. 1294). Springer. https://doi.org/10.1007/978-3-030-63322-6_39

Ivanyuk, V., & Levchenko, K. (2020). Intelligent methods for predicting financial time series. In R. Silhavy, P. Silhavy, & Z. Prokopova (Eds.), *Software engineering perspectives in intelligent systems. CoMeSySo 2020. Advances in intelligent systems and computing* (Vol. 1294). Springer. https://doi.org/10.1007/978-3-030-63322-6_41

Ivanyuk, V., & Soloviev, V. (2019). Efficiency of neural networks in forecasting problems. In *Proceedings of 2019 12th International Conference Management of Large-Scale System Development; MLSD 2019*. https://doi.org/10.1109/MLSD.2019.8911046

Ivanyuk, V., Sunchalin, A., & Sunchalina, A. (2020). Development of an intelligent ensemble forecasting system. In R. Silhavy, P. Silhavy, & Z. Prokopova (Eds.), *Software engineering perspectives in intelligent systems. CoMeSySo 2020. Advances in intelligent systems and computing* (Vol. 1294). Springer. https://doi.org/10.1007/978-3-030-63322-6_40

Li, X., Zhu, S., Yüksel, S., Dinçer, H., & Ubay, G. G. (2020). Kano-based mapping of innovation strategies for renewable energy alternatives using hybrid interval type-2 fuzzy decision-making approach. *Energy, 211*, 118679.

Lisin, A. (2020). Valuation of the activities of foreign banks in the Russian banking sector. *Orbis, 15*(45), 53–63. http://www.revistaorbis.org.ve/pdf/45/art5.pdf

Liu, Z.-J., Panfilova, E., Mikhaylov, A., & Kurilova, A. (2021). COVID-19 crisis impact on the stability between parties in crowdfunding and crowdsourcing. *Wireless Personal Communications*. https://doi.org/10.1007/s11277-021-08932-z

Liu, Z.-J., Panfilova, E., Mikhaylov, A., & Kurilova, A. (2022). Assessing stability in relationship between parties in crowdfunding and crowdsourcing projects during the COVID-19 crisis. *Journal of Global Information Management, 30*(2), 1.

Melnichuk, A. V., Melnichuk, Y. A., Bocharov, A. V., Stepanova, D. I., & Kurbanov, S. A. (2020). Constitutional foundations of the functioning of local self-government. *Revista Inclusiones., 7*, 337–346.

Meynkhard, A. (2019). Fair market value of bitcoin: Halving effect. *Investment Management and Financial Innovations, 16*(4), 72–85. https://doi.org/10.21511/imfi.16(4).2019.07.

Meynkhard, A. (2020). Effect of bitcoin volatility on altcoins pricing. *Advances in Intelligent Systems and Computing, 1294*, 652–664. https://doi.org/10.1007/978-3-030-63322-6_55

Mikhaylov, A. (2018a). Volatility spillover effect between stock and exchange rate in oil exporting countries. *International Journal of Energy Economics and Policy, 8*(3), 321–326.

Mikhaylov, A. (2018b). Pricing in oil market and using probit model for analysis of stock market effects. *International Journal of Energy Economics and Policy, 8*(2), 69–73.

Mikhaylov, A. (2018c). The yield of the carry trade strategy. *Finance: Theory and Practice, 22*(3), 52–63. https://doi.org/10.26794/2587-5671-2018-22-3-52-63

Mikhaylov, A. (2019). Oil and gas budget revenues in Russia after crisis in 2015. *International Journal of Energy Economics and Policy, 9*(2), 375–380.

Mikhaylov, A. (2020a). Cryptocurrency market analysis from the open innovation perspective. *Journal of Open Innovation: Technology, Market, and Complexity, 6*(4), 197. https://doi.org/10.3390/joitmc6040197

Mikhaylov, A. (2020b). Geothermal energy development in Iceland. *International Journal of Energy Economics and Policy, 10*(4), 31–35.

Mikhaylov, A. (2020c). Cryptocurrency market development: Hurst method. *Finance: Theory and Practice, 24*(3), 81–91.

Mikhaylov, A. (2021a). Development of Friedrich von Hayek's theory of private money and economic implications for digital currencies. *Terra Economicus, 19*(1), 53–62.

Mikhaylov, A. (2021b). Lichens as indicators of atmospheric pollution in urban ecosystems. *Israel Journal of Ecology & Evolution, 67*(1–2), 60–68. https://doi.org/10.1163/22244662-bja10016

Mikhaylov, A. (2022). Efficiency of renewable energy generation in Russia. *Annals of the Brazilian Academy of Sciences, 94*(3).

Mikhaylov, A., Burakov, D., & Didenko, V. (2019). Relationship between oil Price and macroeconomic indicators in Russia. *Finance: Theory and Practice, 23*(2), 105–116. https://doi.org/10.26794/2587-5671-2019-23-2-105-116

Mikhaylov, A., Danish, M. S. S., & Senjyu, T. (2021b). New stage in evolution of cryptocurrency market: Analysis by Hurst method. In H. Dincer & S. Yuksel (Eds.), *Strategic outlook in business and finance innovation: Multidimensional policies for emerging economies* (pp. 35–45. ISBN: 978-1-80043-445-5). Emerald Publishing Limited. https://doi.org/10.1108/978-1-80043-444-820211004

Mikhaylov, A., Sokolinskaya, N., & Nyangarika, A. (2018). Optimal carry trade strategy based on currencies of energy and developed economies. *Journal of Reviews on Global Economics, 7*, 582–592.

Mikhaylov, A., & Tarakanov, S. (2020). Development of Levenberg-Marquardt theoretical approach for electric network. *Journal of Physics: Conference Series, 1515*, 052006.

Mikhaylov, A., Yumashev, A., & Kolpak, E. (2021a). Quality of life, anxiety and depressive disorders in patients with extrasystolic arrhythmia. *Archives of Medical Science, 17*(1). https://doi.org/10.5114/aoms.2020.101359

Moiseev, N., Mikhaylov, A., Varyash, I., & Saqib, A. (2020). Investigating the relation of GDP per capita and corruption index. *Entrepreneurship and Sustainability Issues, 8*(1), 780–794.

Moiseev, N., Sorokin, A., Zvezdina, N., Mikhaylov, A., Khomyakova, L., & Danish, M. S. S. (2021). Credit risk theoretical model on the base of DCC-GARCH in time-varying parameters framework. *Mathematics, 9*, 2423. https://doi.org/10.3390/math9192423

Morkovkin, D. E., Gibadullin, A. A., Kolosova, E. V., Semkina, N. S., & Fasehzoda, I. S. (2020a). Modern transformation of the production base in the conditions of industry 4.0: Problems and prospects. *Journal of Physics: Conference Series, 1515*, 032014.

Morkovkin, D., Lopatkin, D., Sadriddinov, M., Shushunova, T., Gibadullin, A., & Golikova, O. (2020b). Assessment of innovation activity in the countries of the world. *E3S Web of Conferences, 157*, 04015. https://doi.org/10.1051/e3sconf/202015704015

Mukhametov, A., Bekhorashvili, A., Avdeenko, A., & Mikhaylov, A. (2021). The impact of growing legume plants under conditions of Biologization and soil cultivation on Chernozem fertility and productivity of rotation crops. *Legume Research, 44*, 12.

Mutalimov, V., Kovaleva, I., Mikhaylov, A., & Stepanova, D. (2021). Assessing regional growth of small business in Russia. *Entrepreneurial Business and Economics Review, 9*(3), 119–133. https://doi.org/10.15678/EBER.2021.090308

Nie, D., Panfilova, E., Samusenkov, V., & Mikhaylov, A. (2020). E-learning financing models in Russia for sustainable development. *Sustainability, 12*(11), 4412.

Nyangarika, A., Mikhaylov, A., & Richter, U. (2019a). Influence oil price towards economic indicators in Russia. *International Journal of Energy Economics and Policy, 9*(1), 123–129.

Nyangarika, A., Mikhaylov, A., & Richter, U. (2019b). Oil price factors: Forecasting on the base of modified auto-regressive integrated moving average model. *International Journal of Energy Economics and Policy, 9*(1), 149–159.

Nyangarika, A., Mikhaylov, A., & Tang, B.-J. (2018). Correlation of oil prices and gross domestic product in oil producing countries. *International Journal of Energy Economics and Policy, 8*(5), 42–48.

Qiu, D., Dinçer, H., Yüksel, S., & Ubay, G. G. (2020). Multi-faceted analysis of systematic risk-based wind energy investment decisions in E7 economies using modified hybrid modeling with IT2 fuzzy sets. *Energies, 13*(6), 1423.

Radosteva, M., Soloviev, V., Ivanyuk, V., & Tsvirkun, A. (2018). Use of neural network models in market risk management. *Advances in Systems Science and Applications, 18*(2), 53–58. https://doi.org/10.25728/assa.2018.18.2.582

Ranjbar, O., Chang, T., Nel, E., & Gupta, R. (2017). Energy consumption and economic growth nexus in South Africa: Asymmetric frequency domain approach. *Energy Sources, Part B: Economics, Planning, and Policy, 12*(1), 24–31.

Rathnayaka, R. K. T., Seneviratna, D. M. K., & Long, W. (2018). The dynamic relationship between energy consumption and economic growth in China. *Energy Sources, Part B: Economics, Planning, and Policy, 13*(5), 264–268.

Saqib, A., Chan, T.-H., Mikhaylov, A., & Lean, H. H. (2021). Are the responses of sectoral energy imports asymmetric to exchange rate volatilities in Pakistan? Evidence from recent foreign exchange regime. *Frontiers in Energy Research, 9*, 614463. https://doi.org/10.3389/fenrg.2021.614463

Sunchalin, A. M., Kochkarov, R. A., Levchenko, K. G., Kochkarov, A. A., & Ivanyuk, V. A. (2019). Methods of risk management in portfolio theory. *Espacios, 40*(16), 8.

Uandykova, M., Lisin, A., Stepanova, D., Baitenova, L., Mutaliyeva, L., Yuksel, S., & Dincer, H. (2020). The social and legislative principles of counteracting ransomware crime. *Entrepreneurship and Sustainability Issues, 8*(2), 777–798.

Udalov, I. (2021). The transition to renewable energy sources as a threat to resource economies. *International Journal of Energy Economics and Policy, 11*(3), 1–6.

Uyeh, D. D., Asem-Hiablie, S., Park, T., Kim, K.-M., Mikhaylov, A., Woo, S., & Ha, Y. (2021). Could japonica Rice be an alternative variety for increased global food security and climate change mitigation? *Food, 10*, 1869. https://doi.org/10.3390/foods10081869

Varyash, I., Mikhaylov, A., Moiseev, N., & Aleshin, K. (2020). Triple bottom line and corporate social responsibility performance indicators for Russian companies. *Entrepreneurship and Sustainability Issues, 8*(1), 313–331.

Wang, S., Liu, Q., Yuksel, S., & Dincer, H. (2019). Hesitant linguistic term sets-based hybrid analysis for renewable energy investments. *IEEE Access, 7*, 114223–114235.

Yuan, G., Xie, F., Dinçer, H., & Yüksel, S. (2021). The theory of inventive problem solving (TRIZ)-based strategic mapping of green nuclear energy investments with spherical fuzzy group decision-making approach. *International Journal of Energy Research., 45*(8), 12284–12300.

Yüksel, S., Dinçer, H., Çağlayan, Ç., & Uluer, G. S. (2021a). Determining the optimal financial strategies for nuclear energy companies. In *Financial strategies in competitive markets* (pp. 1–16). Springer.

Yüksel, S., Khomyakova, L., & Mikhaylov, A. (2021b). Energy center selection in G7 industry with fuzzy MOORA. In S. Yüksel & H. Dinçer (Eds.), *Handbook of research on strategic management for current energy investments* (pp. 87–106). IGI Global. ISBN13: 9781799883357. https://doi.org/10.4018/978-1-7998-8335-7.ch006

Yüksel, S., Mikhaylov, A., & Ubay, G. G. (2021d). Factors causing delay in the installation of nuclear power plants. In S. Yüksel & H. Dinçer (Eds.), *Strategic approaches to energy management. Contributions to management science*. Springer. https://doi.org/10.1007/978-3-030-76783-9_7

Yüksel, S., Mikhaylov, A., Ubay, G. G., & Uyeh, D. D. (2021c). The role of hydrogen in the Black Sea for the future energy supply security of Turkey. In S. Yüksel & H. Dinçer (Eds.), *Handbook of research on strategic management for current energy investments* (pp. 1–15). IGI Global. https://doi.org/10.4018/978-1-7998-8335-7.ch001

Yumashev, A., & Mikhaylov, A. (2020). Development of polymer film coatings with high adhesion to steel alloys and high wear resistance. *Polymer Composites, 41*(7), 2875–2880.

Yumashev, A., Ślusarczyk, B., Kondrashev, S., & Mikhaylov, A. (2020). Global indicators of sustainable development: Evaluation of the influence of the human development index on consumption and quality of energy. *Energies, 13*, 2768.

Zhao, Q., Cherkasov, S., Avdeenko, A., Kondratenko, L., & Mikhaylov, A. (2021). Integral estimate of the added value of agricultural crops in the synergetic agriculture on the example of vicia faba (Vicia faba L.). *Polish Journal of Environmental Studies, 30*(5), 1–9. https://doi.org/10.15244/pjoes/131949

Zhou, P., Luo, J., Cheng, F., Yüksel, S., & Dinçer, H. (2021). Analysis of risk priorities for renewable energy investment projects using a hybrid IT2 hesitant fuzzy decision-making approach with alpha cuts. *Energy, 224*, 120184.

Digital Activist Movements for Energy Resources: The Case of Greenpeace Turkey

Başak Gezmen

1 Introduction

Information is at a point where everyone can create, access, and use it. In new digital world, non-governmental organizations that operate independently of the state, realize activities by setting agenda for different issues in order to create awareness. In this point, media is the strongest tool when its power and role are considered in forming public opinion. The power of internet and social media in today's society is incontrovertible reality. The awareness-raising studies of NGOs bring together many people gathered around the same idea in the virtual environment. Besides, awareness is strengthened with digital activism, which is the fast arena of the digital world. Therewithal, it is frequently encountered the concepts of running out energy resources, energy saving, climate crisis, ecological balance, and environmental literacy in recent time. With regard of these, Greenpeace Turkey, which is one of the most well-known and active organizations on environmental movements, makes an effort to raise awareness and take action in many areas about energy within the scope of digital activist movements.

In this paper, energy-based tweets on Greenpeace Turkey's twitter account are investigated and evaluated within the framework of the determined periods. In this context, the extinction of energy resources and energy saving, the approaches of NGOs about energy consciousness, which themes on energy were emphasized, and what should be done about it, solution suggestions are assessed in the scope of activist movements.

B. Gezmen (✉)
The School of Communication, İstanbul Medipol University, Istanbul, Turkey
e-mail: bgezmen@medipol.edu.tr

© The Author(s), under exclusive license to Springer Nature Switzerland AG 2022
H. Dinçer, S. Yüksel (eds.), *Sustainability in Energy Business and Finance*,
Contributions to Finance and Accounting,
https://doi.org/10.1007/978-3-030-94051-5_13

2 The Transition from Old Social Movements to New Social Movements

Social movements continue their existence in almost every period of life throughout human history. This situation can be explained by the reality of objection and resistance to the process. The concept of social movement is mostly associated with the modern period. Post-modern term paved the way for the transition to a different, new life and also, the formation of social movements in modern sense. Further, Dacheux explained this situation as the emergence of a dramatic opponent-development which realized with the intersection of the division of the masses and the depression of nation-states' crisis and parliamentary debates. Hence, the new social movements are formed at this point. Nevertheless, these movements, which demonstrate themselves in different fields, are social policy issues such as disarmament, women's rights, sexual minorities, racist and ethnic groups, environment, social security, and housing (Dacheux, 2012). In every period of the history, the ruler and the ruled relationships were inquired within the scope of reconciliation and conflict. Thus, many different groups who would like to send their demand and wishes to the government were emerged. The new social movements have busted with experienced of the changes and the transformations after 1980 that is the term of post the working-class movements and independence movements which are called the old social movements. Further, women movements, environmental movements and peace movements are some of the new social movements. In new term, acceleration of capitalist process and globalization arguments, development of multinational companies, and new technologic developments have begun to bring the concept of democracy, human rights, and freedom into the forefront. In addition to that, they have led to the emergence of the new social movements (Balkaya, 2014).

There are some different points between the old and the new social movements. In the old social movements, there was gathered around a leader who was accepted as a hero or a treacherous opponent by the power. Hence, the leaders were in the center. Otherwise, there are some forms that share the power in the new social movements. Particularly, it is noteworthy that the participants of the new social movements are high-level educated and predominantly young people. On the other hand, the old social movements were based on economic and social problems. However, the new social movements are based on human rights, democratic rights and identity, which were not discussed a lot in the past. Besides, the main emphasis in the new period is shaped around the discourses of diversity and otherness (Ayhan & Baloğlu, 2019).

"Man is a social being by nature." is the starting point of Jürgen Habermas's philosophical thoughts within his own words. In the Habermas's philosophy, the public sphere is defined as the space of free discussion where individuals can leave from their private spaces and can participate to the discussion as equal citizens, and have a say on all kinds of social, cultural, and political issues (Torun, 2020). Furthermore, the theorist stated that the public sphere defines the process of communication, dialog, and negotiation in civil society and the methods and tools used in this process. Moreover, the public sphere is the indispensable key of the democratic and social culture (Ceyhan, 2007). Habermas is a positivist and pragmatic philosopher. Also, he is one of the important representatives of the Frankfurt School. According to the theorist who focused on the dominance of instrumental rationality in industrial societies, dwelling on instrumental rationality rather than goals supported the scientification of politics, and political problems came down to problems related to technical control. However, the field of public discussion and negotiation about social goals have been completely forgotten. As the theorist saw, the criticism is the key path for being an open and non-pressured society. (Tekinalp & Uzun, 2006). Therewithal, Habermas, who presented the thesis of Structural Transformation of the Public Sphere in the 1960s, pointed out in this approach that there is an actual public space where social problems were discussed in the eighteenth century. Habermas denoted that the effectiveness was provided at this point, and he specified that these were places of meeting and discussion. Therewithal, human beings had a chance to convey their views to the society within the framework of the emergence of the press with the invention of the printing press. Nonetheless, the bourgeois class occupied the public sphere with the dominance of the industrial capitalist process. With regard of this, Habermas remarked that the organizing of the media and politics as institutional has caused the collapse of the public sphere. While Jürgen Habermas was idealizing the public sphere, he presented a single and holistic public sphere and did not allow different public sphere (Yaylagül, 2010).

3 Social Engagement in the Digital World: Social Media and Digital Activism

The emergence of the Internet started in the second half of the twentieth century. It is communication and information tool that caused a transformation on societies which became an international network over time. Technology is one of the important infrastructure elements that provide social change. Also, internet provides instantaneity, speed, easiness, and relatively low cost. It is a widespread and effective technology that is rapidly rising all over the world apart from continuing ownership and usage inequalities in certain categories (Dedeoğlu, 2016).

In today's world, information society is one of the common concepts that we met frequently. It is a notion related to the societies that use mostly information and communication technologies (ICT) as the component of economic, political, and cultural life. Furthermore, social practices of them are formed by ICT. While these technologies are developing, they had also shaped their ICT. Moreover, this transformation started to be used in parts of the life such as banking, tourism, health, media, and so on. with accelerating technologic developments in the last decade of the twentieth century (Dedeoğlu, 2016).

Moreover, one of the most important and obvious effects of IT on the life practices of individuals is the continuous increase of internet usage. Due to its capacity, the Internet transforms and also transforms the societies at a dizzying speed. People can access the information as they wanted at any time, every minute. Additionally, they can be informed about the developments, have a good time to spend and chat with the people they would like to be thanks to the internet. On the other hand, shopping activities and banking transactions can also be realized by the internet. The most important feature of the internet is to provide interactive communication. Accordingly, it can ensure the elimination of the hierarchy and information monopoly. Furthermore, it is submitted that Internet allows to form new and interesting communities, to bring together individuals with common interests, to strengthen social networks by eliminating time and space (Gülnar & Balcı, 2011). We constantly encounter the concept of the new media in our lives with the development of new communication technologies. Besides, it can be as those systems that can include mass audiences as individual users, and where individuals can access content or applications at different times, and they can be in interaction (Geray, 2002). Due to digitization, data is easily processed, texts are prepared for word processing, and sounds and images become higher quality. Also, the uniform language makes the content more precise (Dijk, 2018). The traditional media in the pre-internet era are the communication environments of magazines, newspapers, radio, and television. With regard of this, the traditional media is one-sided, because it cannot provide active participation to the reader, listener, and audience. So, connections can only be made by telephone and live broadcasts. At this point, criticism and opinions make late, and returns can provide with reader letters and so on (Dedeoğlu, 2016). On the other hand, the concept of new media has come to the forefront by replacing multimedia since the 1990s. Furthermore, this concept is not multiple but composite. Additionally, it provides individualized connectivity and is variable (Akar, 2010). Therefore, the new media plays an efficient and functional role in the creation of social models in the construction of social structure and production relations with these features (Törenli, 2005).

Social media is adopted as one of the most important possibilities of the digital world. It offers users a very free and participatory environment and provides news information data flow. Thus, social media give an opportunity to interact for users in political, ideological, and cultural fields. Hence, users who can gain different identities in social media that has a fast and fluid structure can easily share their thoughts and feelings.

In past periods, individuals who come together in public spaces such as cafes, parks, and so on that are common areas of sharing and discussing social issues and problematics, creates virtual public spaces in today's digital world. Developments in the fields of technology, transport and communication change and transform the social structures. Also, all kinds of organizations and processes involved in society are also affected (Ayhan & Baloğlu, 2019). Therefore, new communication technologies have enabled individuals to meet and come together. It has a public content when it is defined as communication spaces that produce sociality along with this encounter. There is a situation associated with access to a common space (Timisi, 2003).

Activism is in our daily life since ancient times and is based on social movements. The old social movements have been begun with labor movements of the nineteenth century and have been used for expressing peace movements, feminist movements, minority movements, and local autonomy movements, which have gained momentum since the 1970s. The internet use of different types of activism is called "digital activism." A digital activism is to take action over the internet in a digital environment in the form of the advocacy of a goal, organizing around this goal in order to achieve it, transmitting relevant messages to the masses, lobbying, boycotting, and site blackout. Activists report company and government activities for their goal that is brought together them. Moreover, they provide a wide range of information and knowledge to a wide range of audiences through social media, blogs, podcasts, images, video content, and sharing websites. They try to change the ideas of individuals as opponents or advocators on social and political issues. Therewithal, activists in the digital area have opportunities to express themselves better (Livberber, 2019).

It is estimated that the new media directly contributes to the development of civil rights through the relationship between new media and democracy. Nevertheless, the state plays a very important role in the defense of civil rights of the media, which is seen at the peak of the triangle of civil society democracy, in ensuring the formation of public opinion, in illuminating public opinion and in realizing the flow of information. In this regard, the concepts are gaining importance, such as alternative media, online publishing and so on (Dağtaş, 2007).

Alternative media can be explained as ways to organize, produce, and use outside actual realized system and motivated by different goals and norms, without commercial and public institutions. Alternative media as a stance against the industrial structure of the capitalist system media can be explained as a means of symbolic resistance tool for the media of social movements that offer a different field (Andersson, 2017, p. 92–93). The purpose of alternative media that interested in news is simple: to ensure that these groups have access to the media according to their musts. It means that the media should develop to promote and normalize this type of access, such as employees, sexual minorities, unions, protest groups. Also, managers and senior professionals can make their own news (Atton, 2006, p.11).

Lievrouw remodeled the definition of new media in general terms. According to the definition, an alternative activist uses new communication technologies information and communication works, practices and social arrangements of the new media in order to change sovereign, ordinary, and accepted forms of the society, culture, and policy making or to challenge them. Also, it is highlighted that the new media is used to change things (Lievrouw, 2016). Moreover, the new media, which has an opportunity of interaction on digital platforms, allows content production. In this point, activists preferred the new media platforms that is aimed social and politic change. Besides, many examples of digital activism influence social policies by raising awareness and consciousness (Göksu & Durmuş Bektaş, 2019). Digital activism changes and transforms as technology advances. The activists' goals and method preferences determine the types of activism. According to this, types of activism can be classified as participatory/awareness (advocacy) activism, clicktivism/slacktivism, hacktivism, citizen journalism and resource creation. The be advocator/participatory is to organize social media signature campaigns, media campaigns and so on in order to hear the voice of ideas, protect rights and raise awareness (Turhan, 2017). Furthermore, slacktivism is thought to be beneficial in the sense of mobilizing people by opening hashtags, sharing the same profile photo, sharing black ribbons. However, there are also people who think that the feeling of taking action begins to vanish at this point (Turhan, 2017). In the scope of the new communication technologies, each person can report any event, photograph it, create a video, make their own news as a citizen journalist, and create awareness. On the other hand, hacktivists start some activist movements by capturing the information of institutions and organizations in the digital environment to acquire the facts on behalf of the citizen to give perspective to the unseen faces of events.

The new social movements that emerged in the cultural sphere and are based on identity and are unlike the old social movements different. With the development of the new social movements, digital activists have come out in different fields such as women's movements, peace movements, environmental movements and so on.

The first comprehensive environmentalist action in France was Jean Dorst's successful campaign with slogans like before nature dies for the rescue of the Vanoise National Park in 1965 and the 1969 French Federation of Nature Conservation associations. These associations have entered into struggles to ensure that nature does not disappear. While some groups are looking for ways to protect nature but all nature, some groups are more interested in protecting their environment. The highways, industrial zones have become the target of associations formed by people who will live on these areas and who will be to forced migration from this area because of river pollution (Sımonnet, 1993). In today's societies, the continuous consumption-oriented lifestyle, markets' production method in the form of disposable productions and consumption methods cause not only alienation of people against to the nature, but also, they are reason of running out natural resources due to excessively use. Also, they lead to widespread pollution of accumulated waste. These routine works like everyday activities are actually decisive in environmental degradation (Özdemir, 2019). As environmental problems increased and the point of how to solve these problems became important. Therefore, the environmental studies conducted after 1900 focused on the solution. A notable aspect of these years is that public opinion on environmental quality has reached the highest level in all time. An increasingly visible majority views the environmental issues as serious threats to human well-being. With regard of this, Hardoy et al. focused on the concept of a sustainable city profile for minimizing environmental damage and solutions to some existing environmental problems. It is important to be able to provide healthy, safe environments without anticipating unsustainable demands on cities, natural resources, ecosystems, and global cycles (Cansaran, 2019). Particularly, the changes and transformations after 1980, economic development motives triggered by neoliberal policies caused an atmosphere of anti-environmental industrial development in the business world, public groups, and the state bureaucracy. Besides, the concepts such as green buildings and environmentally friendly mining have started to be heard frequently. As environmental history highlighted by Hughes "the task of environmental history is the study of human relationships through time with the natural communities of which they are part, in order to explain the processes of change that affect that relationship." Hence, it is discoursed the dependence of societies on nature and the approach that some changes occur in nature as well as in individual and social life because of it (Hughes, 2019). On the other hand, environmental movements bear some similarities to other forms of movement such as women's movements and peace movements at many points. However, environmentalist movements settle eco-centrism instead of human centrism by proposing a new model of society. Thus, environmentalist movements oppose the understanding of growth and development because it causes environmental problems to propose an eco-centrist and self-sufficient model of society that limits growth (Balkaya, 2014).

Greenpeace and Friends of the Earth are among the most important NGOs' names in environmental movements today. The work of Greenpeace and Friends of the Earth, which have signed global works in raising awareness decisively. These organizations are active in the areas of nuclear power plant opposition, marine ocean pollution prevention, climate change, ending dependency on fossil fuels, environmental justice, and nature (Balkaya, 2014).

4 Energy Efficiency and Sustainable Energy

Energy occurs in different ways as the ability of an object or system to perform works. It is in our life in the form of thermal energy, electrical energy, mechanical energy, chemical and nuclear energy and thermal energy, light energy. Further, reaching energy resources are the fundamental need in our era. There are two energy groups which are renewable and non-renewable (Özdemir, 2020).

Renewable energy is obtained from the existing energy flow in continuous natural processes. Traditional biomass by burning wood, plants, and other substances in traditional ways and large hydrological energy are traditional renewable energy; wind, solar, wave, ocean, geothermal energy are new renewable energy sources (Zhong et al., 2020; Yüksel et al., 2020; Li et al., 2020; Xie et al., 2021). Non-renewable energy sources are also called fossil energy sources in the form of oil, natural gas, coal. According to some approaches, nuclear power plants are included in non-renewable energy resources and they are included in renewable sources (Sevim, 2019).

Both in the world and in Turkey, raising energy production from domestic and renewable energy sources and increased energy efficiency is a mandatory policy in reliable time of the energy within the scope of supplying uninterrupted and environmental compliance energy (Basa & Pamir, 2014). Energy problem is one of the main annoyances of the industry. Also, the exponential increase in energy consumption and the continuous depletion of fossil beds in nature show us that the sad end is approaching. In a world where everything is limited, unlimited material development is impossible. Except running out of resources, environmentalists debate the integrity of the energy system for many reasons. Unfortunately, the system does not depend upon energy that can renew itself. Otherwise, an unstable economy is formed, and it is dependent on resources that will run out in the long term. On the other hand, the crisis of the energy system is the same as the level of energy dependence of users. The excessively used electricity also reflects consumerism in our daily lives (Sımonnet, 1993). As energy is one of the most important natural resources, it is intertwined with economic growth and environmental problems (Yapraklı, 2013).

5 Non-governmental Organizations (NGOs) and Digital Media

Non-governmental organizations (NGOs) that unite around a common goal and act in accordance with the legal dimensions defined for them. They act not for themselves but primarily in accordance with the interests of societies. NGOs are organizations that perform all these activities without being part of states. Although they are affiliated to the state, NGOs do not act as part of the state. Trade unions, political parties, cooperatives, associations, foundations are among the structures that operate as NGOs. Moreover, NGOs, which provide a bridge between society and the state and strengthen the structure of civil society, work on a voluntary basis to enlighten and inform the public (Akay, 2019). Additionally, NGOs have become a concept that is not separated from the concept of democracy. So, it is considered with democracy concept today. Non-governmental organizations which can be defined as non-profit and non-state institutions that contribute to the development of democracy working for the benefit of society. Hence, they are a cross-section of society that is outside the state in the modern sense. Independent NGOs act in accordance with legal, social, cultural, political, and environmental targets. They organize campaigns provided by lobbying, persuasion and actions, donation, or membership fee. Also, they make activities to mobilize and create awareness. In this context, the media is an indispensable tool for these organizations. Due to the collectivist nature of this platform, both local governments and non-governmental organizations have started to use this digitized structure of social movements with technologic developments and accelerating internet and social media. Forums, blogs, microblogs content groups allow individuals to share ideas very quickly and act integrated around a common goal. A communication environment is created that allows them to cooperate in an organized way (Akay, 2019).

In the last 30–40 years, many environmental problems, especially ozone depletion, global warming, and climate change, have become issues that the global political field and public opinion with sensitivity and concern have focused on (Ataman & Erkmen, 2012). The examples of Tweet content as Greenpeace Turkey Twitter account energy-themed shares were considered based on the Twitter account popularity ranking conducted on 10.08.2021. All content shared with the word "energy" which was searched as a keyword in content shared from Twitter account, is the subject of this research. In this study, it is examined which topics Greenpeace Turkey's energy-related posts focus on. Greenpeace is one of the most effective organizations in environmental movements. In 1971, a group of ecologists, journalists, and hippies united around the same idea as the founders of Greenpeace raised a Greenpeace flag on Phyllis Cormack that is a fishing boat, and they sailed to Amchitka Island. It is accepted the starting point of a great movement and a green and peaceful history. The founders 'goal was to stop nuclear tests conducted by the US Navy. These beginnings of Greenpeace have subsequently evolved into a movement in the international arena. Their main goal is to protect the environment and to create change. Greenpeace Turkey is Turkish field of the group. They always aim at peaceful ways and try to overcome difficulties. Moreover, Greenpeace Turkey consists of people who promise to fulfill the core values of Greenpeace and come together to defend the right to live in a healthy environment.

In this study, it is investigated Greenpeace Turkey's tweets on shared of energy under the popular category on 10.08.2021.

In the posts on May 26, 2019, May 27, 2019, and May 30, 2019, and June 7, 2019, under the popular subtitle, it was emphasized that the energy consumed by a stadium with 55 thousand people in a match is equal to the annual consumption of 164 households and 55 thousand people stadium causes 3600 tons of carbon emissions. Hundreds of people who came to Istanbul and visited the Rainbow Warrior ship started a campaign to pioneer football clubs for a clean future with the slogans "Shoulder to shoulder for the sun," "Look alive," "Let make the stadiums light up with solar energy," and "Better energy brings goals."

On January 9, 2020, and January 18, 2010, a podcast was published in which energy expert Ceren Ayas assessed Australian fires in the second post. On June 15, 2020, the third post was about the share of fossil fuels in energy consumption. According to the REN21 report, the share of fossil fuels in ultimate energy consumption has declined from 80.3% in 2009 to only 80.2% in 2019.

In the post on November 22, 2019, the news was given that the oldest 15 coal thermal power plants of Turkey will pollute the air more than 2.5 years. It was stated that more than 100,000 people who say "Leave coal" will continue to fight for energy with the call of "Join us" on the same date. In the fourth post on fifth of May 2020, it drew attention to the necessity of giving up fossil fuels in electricity generation, giving weight to bicycle use in cities and switching to renewable energy sources in order to mitigate the climate crisis with the "The future is on the bike." In the fifth post on February 2, 2020, it is stated that the need to close coal-fired thermal power plants in Turkey. Additionally, it was included that wind and solar energy production should be increased in Europe. Furthermore, the sixth post on October 8, 2020, is for coal-based energy production. It was stated that Kemerköy Region is in the fourth and Afşin and Elbistan are in the fifth place in air pollution caused by thermal power plants worldwide. Next, the seventh post on December 30, 2020, was aimed at the necessity to increase the potential of renewable energy. In the eighth post on April 30, 2020, it was again on the subject of struggle for renewable energy.

A call was made to support the campaign by using expressions such as "it is time to carry our struggle with special cloth bags on our shoulder for the 25th Anniversary." "Click, choose your campaign, get one of the limited-edition bags;" in the same post, it was emphasized that the banner of "Coal Kill Our Lives" hung by a hot air balloon flying over the area where 6000 trees were cut down for the coal-fired thermal power plant in Soma, Manisa in January 2016. Therewithal, it was denoted that the struggle in Soma resulted in victory. In the nineth post on May 1, 2019, it was shared the content supported by the video contents which included villagers who autographed and tried to protect their lands on the verge of extinction due to expanding mines. Accordingly, "Call and Sign" shares are made to The Ministry of Energy and Natural Resources.

In tenth post on March 10, 2019, it was supported the share about the video content for the signature campaign on the same issue related to the mother and child crying out because her child is not healed in Dilovası.

In eleventh post on March 6, 2019, it was remarked that people thought of pollution, disease, and death in the signature campaign when they heard about thermal power plant. In twelfth shared video of the repeated signature campaign on February 20, 2019, it was set the agenda that Çanakkale Çan Yaya Village is referred to as the area left to die because of two coal-fired thermal power plants.

The thirteenth post on April 2, 2020, was about the fight against coal-fired thermal power plants. In this regard, the statements of Onur Akgül, who is in charge of Climate and Energy Project are included.

2020 Climate Transparency Report in fourteenth post on18th of November 2020 is investigated. It is advocated the requirement to stop support for fossil fuels. In the fifteenth post on June 1, 2020, it was launched by Greek Minister of Environment and Energy Kostis Hacidakis the "Greece" where are no disposable plastic products throughout the country.

In the sixteenth post on February 15 it was included that the Spanish Minister of Energy, Teresa Ribera announced that seven nuclear power plants across the country will be shut down by the decision of the government until 2035.

In the post on June 11, 2020, the requirement to unplug the computers after they are turned off and computers consume much lower energy in sleep mode is supported by the slogan "Put your computer to sleep when you did not work."

On March 22, 2019, it is stated that the use of coal for energy generation in Finland will be completely banned from May 1, 2029.

In the post on seventh of November 2019, it is involved that the residents of Zonguldak, Çanakkale and Afşin Elbistan applied to the Ministry of Energy and Natural Resources for the decommissioning of 15 coal-fired thermal power plants operating without a chimney filter. Also, hashtags shared in the content, such as "The Clean Air is The Right."

On July 7, 2019, Ajax's stadium Johan Cruff Arena generates electricity with 4200 solar panels on its roof of. Moreover, it is requested to support the lighting of the stadiums with solar energy to the statements that this energy is stored in the batteries of 148 electric vehicles.

In the posts on February 12, 2019, it was discussed that click and share your opinion about whether the permanent daylight-saving time application, which started in October 2016, can save energy or not.

On June 29, 2019, the slogans of "Take action, The stadiums light up with solar energy" were supported.

On February 19, 2019, the slogan of "Leave Coal" is supported in the video that is about the Hasibe Koç's life struggle in Ceyhan, Adana.

There is a post is for the permanent daylight-saving time application on February 8, 2019. "Click and join our one-question survey."

In the post for May 1, 2020, Labor Day, it is stated that there are about 40 thousand people working in coal mines in Turkey, and it is pointed out that coal workers do not have to work under risk in dirty and old energy production.

On January 2, 2020, the news of the decommissioning of five active coal power plants on the grounds that they did not install a chimney filter in the given time is discussed with the rhetoric of victory. It is expressed that this is the success of hundreds of thousands of people who say "Leave coal." Furthermore, it is declared that the struggle will also continue in 2020.

In the post on September 21, 2020, it is announced that Turkey is accelerating studies of nuclear energy while the world is phasing out nuclear energy reactors. Also, the second power plant planned to be built in Sinop after Akkuyu Nuclear Power Plant that is the first power plant in Mersin, Turkey and it was approved. Therefore, the statements of "Our future is in danger" draw attention.

6 Conclusion: Discussion and Suggestions

New media which provides social interaction between physically distant individuals from each other in the digital world, allows individuals to ensure the collectivity required by public interaction. Due to its influence on the transformation of time and place, new communication technologies and the internet emerged as a virtual place that is a social participation place. In the digital world, which causes a more participatory life, the internet creates a new public sphere arena.

It is ensured that everyone is open to participation, the features of interaction, the ability to express their thoughts freely, and all topics in conflict put forward by discussing. In this context, some approaches positioned the mass communication tools as an important for the democracy in the digital world. However, the non-egalitarian structure as a requirement of the capitalist system and political economy approaches should not be ignored in here.

In this virtual sphere where everyone can gather and act around a certain purpose, NGOs that are non-profit organizations that produce social responsibility projects and aim to support social issues, organize activities to raise awareness and mobilize for the problems. Hence, traditional old social movements gain momentum with digital activism as the realized form in the digital arena.

Nowadays, industrialization, rapid population growth, chemicals polluting the environment have become the magnitude that can be harmful to living. Furthermore, the ecological balance has deteriorated, climate crises have begun to increase, and energy resources have been depleted as a result of excess consumption. In this context, it is tried to prevent the environment-oriented destruction within the scope of environmental movements by focusing on the issues such as energy saving, protecting energy resources, and creating energy awareness. Besides, Greenpeace Turkey realized quite important activities in the framework of environmental movements for the protection of energy resources which is the main subject of this study. In the context of the research, Greenpeace Turkey's energy content on the axis of digital activist movements was evaluated. With regard of these, the themes on energy within the scope of the examined contents are mainly energy saving, the share of fossil fuels in energy consumption, coal-fired power plants, air pollution, and the climate crisis. Moreover, digital activist movements are supported by signature campaigns, announcements, sharing of hashtag posts, and click-through campaigns. With the NGOs have started to use the digital world effectively and successfully, environmental movements in social media have accelerated. Successful campaigns which are carried out by people who unite within the framework of the same view by raising awareness on the common problems are also shared in the tweets. In this context, creating energy consciousness, which is the most important issue of our time, should be tried to be thought from a young age through educational programs. Additionally, it should continuously be kept on the agenda with activist movements of the NGOs that are particularly active in social media. Also, the competency of energy literacy should be learned through solution-oriented, influential movements and scientific studies.

References

Akar, E. (2010). *Sosyal Medya Pazarlaması [Social media marketing].* Ankara.

Akay, R. A. (2019). Sivil Toplum Kuruluşları ve Dijital Aktivizm *[Non-governmental Organizations and Digital Activism].* In H. Taşkıran Boztepe & M. Mengüiçinde (Eds.), *Dijital Aktivizm Üzerine [as cited in About Digital Activism]* (p. s. 139–178). Der Yayınları.

Andersson, L. (2017). Neoliberal Söyleme Karşı Alternatif Medyanın Eleştirel Gücünün Alternatifsiziği [The non-alternativeness of the critical power of alternative media against neoliberal discourse]. F. Aydoğaniçinde, as cited in *Yeni Medya Kuramcılarından Yeni Medya Kuramları [as cited in New Media Theories from New Media Theorists]* (s. 85–110). Der Yayınları.

Ataman, B., & Erkmen, Ö. (2012). Liberal Basın Teorisi ve Ulusal Basında Çevre Haberleri *[The theory of liberal press and environmental news in national press].* In S. Kırlı & Ü. Sarıiçinde (Eds.), *Medya ve Gündelik Yaşamda Çevre Gerçeği [as cited in Media and The Fact of Environment in Daily Life]* (pp. 3–48). Aya Kitap.

Atton, C. (2006). *Alternative Media.* Sage Publications.

Ayhan, B., & Baloğlu, E. (2019). Yeni Toplumsal Hareketler ve Sosyal Medya *[New social movements and social media].* In T. Livberger & S. Tiryakiiçinde (Eds.), *Sosyal Medya Çalışmaları Kuram, Söylem, Pratik [as cited in Social Media Studies, Theory, Discourse, Practice]* (pp. 263–286). Literatürk.

Balkaya, F. (2014Cilt: 3. Sayı:5). Yeni Toplumsal Hareket Çeşidi Olarak Çevreci Hareketler [Environmentalist movemetns as the variety of new social movements]. GJEBS Global Journal of Economics and Business Studies, 32–42.

Basa, N., & Pamir, N. (2014). *Enerji ve Hukuk Sempozyumu Sonuç Bildirgesi Temel Saptamalar ve Sorunlar Çözüm Önerileri [Final declaration of energy and law symposium the fundamental findings and suggestions for solution of problems]*. Türkiye Barolar Birliği.

Cansaran, D. (2019). *Sürdürülebilir Kentler için Çevre Bilinci [Environmental consciousnees for sustainable cities]*. Gece Kitaplığı.

Ceyhan, Ç. (2007). Avrupa Birliği Çevre Politikalarında Katılım Hakkı: Aarhus Sözleşmesi *[Right to participate for environmental policies in European union: Aarhus convention]*. In N. Türkoğlu & M. C. Şimşekiçinde (Eds.), *Medya Okuryazarlığı [as cited in Media Literacy]* (pp. 50–67). Kalemus Yayınları.

Dacheux, E. (2012). *Kamusal Alan [Public Sphere]*. Ayrıntı Yayınları.

Dağtaş, E. (2007). Yurttaşlık haklarının inşası Ekseninde Eskişehir Yerel Basınından Online (Çevrimiçi)Bir Örnek: Midas Gazetesi *[An online example from Eskişehir local press on the axis of civil rights building: Midas newspaper]*. M. Binarkiçinde, *Yeni Medya Çalışmaları [as cited in New Media Studies]* (s. 97–145). Dipnot Yayınları.

Dedeoğlu, G. (2016). *Teknoloji, İletişim, Yeni Medya ve Etik [Technology, communication, new media and ethics]*. Sentez Yayıncılık.

Dijk, J. V. (2018). *Ağ Toplumu Alternatif Medya ve Toplumsal Hareketler [Network society, alternative media, and social movements]*. Epsilon Yayınevi.

Geray, H. (2002). *İletişim ve Teknoloji Uluslararası Birikim Düzeninde Yeni Medya Politikaları [Communication and technology, new media politics on the pattern of international accumulation]*. Ütopya Yayınları.

Göksu, N. F., & Durmuş Bektaş, S. (2019). Sosyal Politikalar Bağlamında Dijital Aktivizm *[Digital activism in the scope of social politics]*. In H. Taşkıran Boztepe & M. (Eds.), *Mengü içinde, Dijital Aktivizm Üzerine [as cited in About Digital Activism]* (pp. 33–62). Der Yayınları.

Gülnar, B., & Balcı, Ş. (2011). *Yeni Medya ve Kültürleşen Toplum [New media and acculturation of the society]*. Literatürk.

Hughes, J. (2019). *Çevresel Tarih Nedir?[What is environmental history?]*. Tarih Vakfi Yurt Yayınları.

Li, X., Zhu, S., Yüksel, S., Dinçer, H., & Ubay, G. G. (2020). Kano-based mapping of innovation strategies for renewable energy alternatives using hybrid interval type-2 fuzzy decision-making approach. *Energy, 211*, 118679.

Lievrouw, L. (2016). *Alternatif ve Aktivist Yeni Medya [Alternative and activist new media]*. Epsilon Yayıncılık.

Livberber, T. (2019). Kendini İfade Etmenin Dijital Hali: Sosyal Medyada Aktivizm *[Digital form of expressing yourself: Activism in social media]*. In T. Livberber & S. Tiryakiiçinde (Eds.), *Sosyal Medya Çalışmaları, Kuram, Söylem, Pratik [as cited in Social Media Studies, Theory, Discourse, Practice]* (pp. 287–304). Literatürl.

Özdemir, O. (2019). Gündelik Yaşam Pratikleri ve Çevre Eğitimi *[Practice of daily life and environmental education]*. In S. Yalmancı Gürbüzoğlu & S. Aydıniçinde (Eds.), *Çevre Etiği Temel İlkeleri ve Eğitimi [as cited in Basic Principles and Education of Environment Ethics]* (pp. 23–37). Nobel Yayınları.

Özdemir, Y. (2020). *Türkiye'nin Enerji Stratejisi [Energy strategies of Turkey]*. Nobel Yayınları.

Sevim, C. (2019). *Enerji Teknolojileri ve Enerji Yatırım Projeleri [Energy technologies and energy investment projects]*. Seçkin Yayınları.

Sımonnet, D. (1993). *Çevrecilik [Environmentalism]*. İletişim Yayınları.

Tekinalp, Ş., & Uzun, R. (2006). *İletişim Araştırma ve Kuramları [Communication research and theories]* (Vol. 2006).

Timisi, N. (2003). *Yeni İletişim Teknolojileri ve Demokrasi [New communication technologies and democracy]*. Dost Yayınları.

Törenli, N. (2005). *Bilişim Teknolojileri Temelinde Hbaer Medyasının Yeniden Biçimlenişi: Yeni Medya, Yeni İletişim Ortamı [Reshaping of news media on the basis of information technologies: new media, new communication environment]*. Bilim ve Sanat Yayınları.

Torun, T. (2020). Jürgen Habermas'ın Kamusal Alan Kavrayışı: Rasyonel Politik İrade Oluşumu *[Jürgen Habermas's conception of the public sphere: The formation of rational political will]. Kaygı*, 220–238.

Turhan, D. G. (2017/1 Sayı: 26). Dijital Aktivizm *[Digital activism]. Süleymen Demirel Üniversitesi Sosyal Bilimler Enstitüsü Dergisi*, 26–44.

Xie, Y., Zhou, Y., Peng, Y., Dinçer, H., Yüksel, S., & an Xiang, P. (2021). An extended pythagorean fuzzy approach to group decision-making with incomplete preferences for analyzing balanced scorecard-based renewable energy investments. *IEEE Access, 9*, 43020–43035.

Yapraklı, S. (2013). *Enerjiye Dayalı Büyüme [The growth based on energy]*. Beta Yayınları.

Yaylagül, L. (2010). *Kitle İletişim Kuramları Egemen ve Eleştirel Yaklaşımlar [Mass Communication Theories, Sovereign and Critical Approaches]*. Dipnot Yayınları.

Yüksel, S., Dinçer, H., & Uluer, G. S. (2020). The role of technological development on renewable energy usage: An econometric analysis for G7 countries. In *Handbook of research on sustainable supply chain management for the global economy* (pp. 136–153). IGI Global.

Zhong, J., Hu, X., Yüksel, S., Dinçer, H., & Ubay, G. G. (2020). Analyzing the investments strategies for renewable energies based on multi-criteria decision model. *IEEE Access, 8*, 118818–118840.

The Stability of Financial Institutions and Counterparties

Zaffar Ahmed Shaikh and Nikita Makarichev

1 Introduction

The extension of insurance coverage to legal entities will be a stimulating factor for such lenders to accumulate financial resources in the banking sector of the Russian Federation. Expanding the scope of insurance will require a review of the current model of the energy risk insurance system, which currently includes only those banks whose license provides for the right to insure energy risks. On the other hand, extending insurance coverage to legal entities, while maintaining the current conditions for mandatory participation in the energy risk insurance system may contribute to the outflow of funds of legal entities from credit institutions that are not participants in the energy risk insurance system. On the other hand, it may be possible to increase the number of employees who are not participants in the energy risk insurance system, which may lead to the formation of a risky business model for such credit institutions.

This creates a moral hazard that depends on the effectiveness of market discipline. It is assumed that an effective market discipline that reduces the degree of moral hazard is a situation in which the project carefully monitors the financial condition of the bank, and, if there is a risk of loss of funds, begins to demand an increase in the deposit rate, or withdraws money from this bank altogether.

Z. A. Shaikh (✉)
Faculty of Computing Sciences & Information Technology, Benazir Bhutto Shaheed University, Karachi, Pakistan
e-mail: zashaikh@bbsul.edu.pk

N. Makarichev
Financial Research Institute of the Ministry of Finance of the Russian Federation, Moscow, Russia

© The Author(s), under exclusive license to Springer Nature Switzerland AG 2022

H. Dinçer, S. Yüksel (eds.), *Sustainability in Energy Business and Fincnce*, Contributions to Finance and Accounting,
https://doi.org/10.1007/978-3-030-94051-5_14

Thus, there is a change in the structure of the bank's financial assets, which reduces its stability. The main thing in this mechanism is the close attention of the project to the bank. Now, since the project already knows that its money is protected, this mechanism does not work, which leads to certain problems and imposes additional burden and responsibility on the control bodies.

It is worth noting that what was in operation 15–20 years ago is no longer suitable for functioning at the present time. That is why it is necessary to make drastic changes in the processes of formation and functioning of the system, which must necessarily have legislative consolidation.

So, there is no need to "reinvent the wheel," since there are many types and interpretations of energy risk insurance systems that are successfully used by other countries, it is only necessary to shift this experience to Russian conditions. It is these urgent changes that will be discussed in the next chapter of this study.

2 Literature Review

According to the data of the Energy Risk Insurance Agency, in 2019, the number of participants in energy risk insurance amounted to 723 banks. By 2020, this number has significantly decreased—to 696 (a decrease of 27 participants in the country compared to the previous period), including existing banks licensed to work with individuals in terms of attracting and placing deposits—372; existing credit institutions that previously accepted energy risks, but currently lost the right to attract funds from individuals –6; banks that have a license to work with individuals in terms of attracting and placing deposits—, in the process of liquidation—318 (Bhuiyan et al., 2021; Dong et al., 2021, Mikhaylov, 2021b; Liu et al., 2021, 2022; Saqib et al., 2021; Radosteva et al., 2018; Ranjbar et al., 2017; Rathnayaka et al., 2018; Sunchalin et al., 2019; Uandykova et al., 2020; Udalov, 2021; Dinçer et al., 2020a, b; Yuksel et al., 2021a, b, c; Dorofeev, 2020; Qiu et al., 2020; Mukhametov et al., 2021; Candila et al., 2021; Zhou et al., 2021).

Banks that previously accepted energy risks but lost the right to attract funds from individuals to deposit accounts include Asia-Invest Bank, State Specialized Russian Export-Import Bank, PROMSVYAZINVEST, Synergy, JSC settlement non-bank credit organization "KHOLMSK," JSC settlement non-bank credit organization "Narat."

The Agency is working diligently to improve the efficiency of finding and returning bank assets that were withdrawn abroad by unscrupulous owners. The number of insured events during the period from 2016 to 2020 tends to fall. This was due to the revocation of licenses from many bankrupt banks, a reduction in the contribution rate during COVID-19 pandemic, outflow of foreign currency deposits from Russian banks (Mikhaylov, 2018c; Mikhaylov et al., 2019; Melnichuk et al., 2020; Nie et al., 2020; Moiseev et al., 2021; Grilli et al., 2021).

Commercial banks regularly pay insurance premiums in the amount of 0.1% of all deposits to energy agencies. Thus, customers do not personally make an additional payment for deposit insurance, and this obligation is performed by the bank at the basic, additional or increased additional rate based on the current legislation (Dayong et al., 2020; Mikhaylov et al., 2018; Nyangarika et al., 2018, 2019a, b; Danish et al., 2020, 2021; Fang et al., 2021; Li et al., 2020; Du et al., 2020; Lisin, 2020; An et al., 2021; Ivanyuk & Berzin, 2020; Ivanyuk & Levchenko, 2020; Yuan et al., 2021; Ivanyuk et al., 2020; Ivanyuk, 2018; Ivanyuk & Soloviev, 2019; Uyeh et al., 2021; Wang et al., 2019; Dinçer et al., 2019).

2020 as a whole was a difficult year for the banking system. Including in the segment of term deposits of citizens. A number of factors provoked citizens to withdraw their savings from credit institutions. Especially during self-isolation. In particular, in March–May 2020. Although, the dynamics even during this period were different, if we evaluate the results of each month relative to the previous one (An et al., 2019a, b, 2020a, b, c; Mikhaylov, 2019, 2021a; Mikhaylov & Tarakanov, 2020; Moiseev et al., 2020; Gura et al., 2020; Dooyum et al., 2020; Mikhaylov et al., 2020a, b, c, 2021a; Varyash et al., 2020; Zhao et al., 2021; An & Mikhaylov, 2020; Alwaelya et al., 2021; Yumashev & Mikhaylov, 2020; Yumashev et al., 2020; Mikhaylov et al., 2021b; Mutalimov et al., 2021; Morkovkin et al., 2020a, b; An & Mikhaylov, 2021).

At present, the banking system of our country is relatively stable in its development, primarily due to the functioning of the energy risk insurance system. During periods of multiple Since the practical implementation of the system, we have managed to avoid the occurrence of numerous cases of mass withdrawal of funds from deposit accounts by projects, as was the case in the 90s of the last century. Thus, we can conclude that the energy risk insurance system is quite successfully coping with the role of ensuring the stability of the country's banking system (Denisova et al., 2019; Nyangarika et al., 2019a, b; Huang et al., 2021a, b; Mikhaylov, 2018a, b, 2022; Meynkhard, 2019, 2020; Mikhaylov et al., 2019; Conteh et al., 2021).

The situation in the economy, falling incomes due to the coronavirus and low interest rates will continue to negatively affect the dynamics of energy risks in the medium term. At the same time, in the second half of 2021, the situation may start to change for the better, but the real growth rate of energy risks by the end of 2021 is likely to be at the level or slightly worse than the result of 2020.

Thus, in 2019, the energy risk insurance system had a positive impact on the market of bank energy risks, contributed to maintaining the positive trends in the field of bank savings of the population that have developed in recent years. In 2021, there is a trend and prospects for expanding the energy risk insurance system, which will increase the inflow of energy risks to banks at the end of the scope of energy risk insurance and the activities of the energy agency allow us to determine the procedure for compensation in the event of an insurance event. a case for all subjects of the process. Therefore, the energy risk insurance system needs to be further developed, as it is an effective tool that helps protect the interests of projects and ensure the relatively stable operation of banks in conditions of economic instability in the global economy.

3 Methodology

As of June 30, 2020, 53 banks were accredited by the Agency to participate in competitions to select agent banks for payment of refunds. Structural divisions of these banks are located in all regions of the Russian Federation, which allows the vast majority of projects to receive compensation at their place of residence. The number of accredited agent banks includes the largest banks in terms of attracted deposits of individuals: Sberbank PJSC, VTB Bank (PJSC), Rosselkhozbank JSC, GPB Bank (JSC) and Otkritie FC Bank PJSC. In the first half of 2020, the Agency will test a new digital a service that allows accepting payment applications in electronic form and paying insurance compensation through remote service channels of the agent bank. It was used by more than 1.3 thousand projects of NVK Bank JSC, which were paid about 520 million rubles through Sberbank Online (an online service of Sberbank PJSC). In addition, more than 1.7 thousand projects in the reporting period received information about the amount of compensation due and paid through the electronic service.

In the reporting period, the results of inspections of banks conducted in 2019 with the participation of the Agency were also summed up. Generalized data on the results of inspections were sent to the Bank of Russia. In general, the results of inspections of banks in 2019 allow us to conclude that banks participating in energy risk insurance comply with the requirements of the Law on Energy Risk Insurance in terms of fulfilling their duties. It should also be noted that deposit insurance takes place automatically when it is opened in a bank participating in the energy risk insurance system (energy risk insurance). The list of participating banks and banks excluded from the insurance system is published on the official website of the Energy Agency. First of all, energy risk insurance includes such large financial and credit organizations as Sberbank of Russia, VTB Bank, ALFA-BANK, and Gazprombank. The amount of insurance liability of the energy agency.

At the end of 2020, the deposit market in Russia showed rather weak dynamics. According to the Bank of Russia, the increase in household energy risks in real terms over the past year was only +4.2%, which is more than 2 times less than in 2019 (+9.7%). For comparison, the last time the growth rate was lower was in 2014 (−2.5%).

It is worth noting that in 2020, due to the depreciation of the ruble, the dynamics of deposits in nominal terms was quite good. The volume of energy risks increased by 8% (2.4 trillion rubles), which is not much different from the dynamics of recent years. Thus, almost half of the increase in nominal terms is a currency revaluation (Dinçer et al., 2020a, b; Girma et al., 2007; Haghi et al., 2018; He et al., 2019; Hoegen et al., 2018; Hong et al., 2015; Koengkan et al., 2020; Lam & Law, 2016, 2018.

The situation in the economy, falling incomes due to the coronavirus and low interest rates will continue to negatively affect the dynamics of energy risks in the medium term. At the same time, in the second half of 2021, the situation may start to change for the better, but the real growth rate of energy risks by the end of 2021 is likely to be at the level or slightly worse than the result of 2020.

4 Analysis Results

Leading Russian experts and scientists identify the following main problems related to the formation and functioning of the energy risk insurance system:

1. A fixed amount of state guarantees established at the state level (since the form of the Energy Risk Insurance Agency is state-owned), which is most often unfair in calculation and insufficient to meet the interests of projects.
2. The system of contributions to the Mandatory Energy Risk Insurance Fund is outdated. All banks deduct contributions at a single basic percentage set by the Central Bank of the Russian Federation using the flat scale method, without taking into account the level of risks of each of them.
3. The system of energy risk insurance in our country, unlike the world experience of using this system, covers a limited range of objects of insurance protection.
4. Insufficient public awareness of the operation of the energy risk insurance system and hidden information about the actual reliability of many banks.
5. Fraudulent actions in the field of energy risk insurance system. For the purpose of intentionally obtaining an illegal insurance refund.
6. A very significant problem today is the policy of the Central Bank of the Russian Federation aimed at "improving" the banking sector.

The Fund's insufficiency necessitates the use of borrowed funds by the Energy Risk Insurance Agency, which is always the beginning of the development of an even deeper and more difficult problem to solve. This is evidenced by the current statistics and forecast data determined for the future, which were discussed in more detail in the previous paragraph.

Taking into account international experience, it is advisable to expand the range of persons (entities) covered by the protection provided by the energy risk insurance system by including legal entities 'accounts and deposits in the system. The extension of insurance coverage to legal entities will be a stimulating factor for such lenders to accumulate financial resources in the banking sector of the Russian Federation.

We believe that funds of legal entities whose activities are related to the provision of financial services and are based, in particular, on a professional assessment of the stability of financial institutions and counterparties should be excluded from the scope of energy risk insurance.

5 Conclusions and Discussion

In this regard, it is proposed not to extend insurance coverage to bank accounts (deposits) of credit institutions; professional participants in the securities market; trade organizers; clearing organizations; microfinance organizations; consumer credit cooperatives; insurance organizations; insurance brokers; mutual insurance companies; non-state pension funds; management companies of investment funds (mutual investment funds) and non-state pension funds; specialized depositories of investment funds (mutual investment funds). private pension funds; pawnshops; leasing companies.

It is also not advisable to insure funds of the federal budget, funds of budgets of constituent entities of the Russian Federation and local budgets, funds of state and other extra-budgetary funds placed in credit institutions, with the exception of funds of state and municipal social institutions (for example, educational and medical institutions). It is debatable to include "funds in settlements" (transfers without opening an account, letters of credit, etc.) in the perimeter of the energy risk insurance system, taking into account the accumulation of such funds in bank accounts. Funds for a relatively short period of time. The inclusion of bank accounts (energy risks) of individuals in precious metals in the insurance coverage period seems impractical due to the predominantly investment nature of such accounts and, as a consequence, their use by investors who are able to assess certain risks associated with the dependence of profitability on such accounts (deposits) on market quotations for metal placed on accounts (deposits).

In addition, it does not seem appropriate to include individual legal entities in the insurance perimeter of accounts (deposits), for which the limit of insurance compensation is generally insignificant. One of the key tasks assigned to energy risk insurance is to protect the funds placed by citizens themselves in banks. In many countries, there is a system for protecting the financial condition and interests of the population, which is perhaps the most important social task. Energy risk insurance is mandatory in any member State of the European Union. As an example, energy risk insurance operates on the territory of Brazil, the USA, Japan, as well as on the territory of the CIS countries-Armenia, Ukraine, Kazakhstan, and others. In general, it is possible to classify existing energy risk insurances in the world according to numerous criteria. The system for insuring monetary energy risks of citizens became a necessary step in connection with the default of 1998. Just in 1998, the state, represented by the government, came to understand that the state really needs and needs such a mechanism, with the help of which it is possible to minimize any negative consequences among banking institutions. Systems with financing are based on the fact that specialized funds are created for payments of deductions for insurance. Funds are formed using regular contributions made by participating banks. Such a system has a fruitful effect on increasing trust, and in the event of an insured event, such a system also accelerates the transfer of funds as compensation.

As for the system without financing, here the funds needed for compensation can only be found if necessary, such as the bankruptcy of a banking institution.

This is a less preferable system, because in the event of a crisis, many banks come under attack, and it is very difficult to collect the necessary amount. Also, fundraising in this system is a long process, which causes panic among numerous projects.

The goals of energy risk insurance based on the rapid elimination of the crisis and its consequences, as well as the formation of a stable system, cannot be achieved.

As of June 30, 2020, the energy risk insurance system included 704 credit organizations, including 353 existing credit organizations that have the right to open new accounts and accept funds of individuals for energy risks; 6 existing credit organizations that have lost the right to open new accounts and accept funds of individuals for energy risks; 345 credit organizations that are in the process of bankruptcy proceedings (liquidation).

The number of accredited agent banks includes the largest banks in terms of attracted deposits of individuals: Sberbank PJSC, VTB Bank (PJSC), Rosselkhozbank JSC, GPB Bank (JSC), and Otkritie FC Bank PJSC.

References

Alwaelya, S. A., Yousif, N. B. A., & Mikhaylov, A. (2021). Emotional development in preschoolers and socialization. *Early Child Development and Care, 190*(1717480).

An, J., & Mikhaylov, A. (2021). Analysis of energy projects financial efficiency and renewable energy generation in Russia. *Finance: Theory and Practice, 25*(5), 1–11.

An, J., & Mikhaylov, A. (2020). Russian energy projects in South Africa. *Journal of Energy in Southern Africa, 31*(3).

An, J., Mikhaylov, A., & Jung, S.-U. (2020a). The strategy of South Korea in the global oil market. *Energies, 13*(10), 2491.

An, J., Mikhaylov, A., & Kim, K. (2020b). Machine learning approach in heterogeneous group of algorithms for transport safety-critical system. *Applied Sciences, 10*(8), 2670.

An, J., Mikhaylov, A., & Moiseev, N. (2019b). Oil price predictors: Machine learning approach. *International Journal of Energy Economics and Policy, 9*(5), 1–6.

An, J., Mikhaylov, A., & Sokolinskaya, N. (2019a). Machine learning in economic planning: Ensembles of algorithms. *Journal of Physics: Conference Series, 1353*, 012126.

An, J., Mikhaylov, A., & Jung, S.-U. (2021). A linear programming approach for robust network revenue management in the airline industry. *Journal of Air Transport Management, 91*, 101979.

An, J., Mikhaylov, A., & Richter, U. H. (2020c). Trade war effects: Evidence from sectors of energy and resources in Africa. *Heliyon, 6*, e05693. https://doi.org/10.1016/j.heliyon.2020.e05693

Bhuiyan, M. A., An, J., Mikhaylov, A., Moiseev, N., & Danish, M. S. S. (2021). Renewable energy deployment and COVID-19 measures for sustainable development. *Sustainability, 13*, 4418. https://doi.org/10.3390/su13084418

Candila, V., Maximov, D., Mikhaylov, A., Moiseev, N., Senjyu, T., & Tryndina, N. (2021). On the relationship between oil and exchange rates of oil-exporting and oil-importing countries: From the Great Recession period to the Covid-19 era. *Energies, 14*, 8065.

Conteh, F., Takahashi, H., Hemeida, A. M., Krishnan, N., Mikhaylov, A., & Senjyu, T. (2021). Hybrid grid-connected renewable power generation for sustainable electricity supply in Sierra Leone: Case study Lungi Town. *Sustainability, 13*(18), 11435.

Danish, M. S. S., Bhattacharya, A., Stepanova, D., Mikhaylov, A., Grilli, M. L., Khosravy, M., & Senjyu, T. (2020). A systematic review of metal oxide applications for energy and environmental sustainability. *Metals, 10*(12), 1604.

Danish, M. S. S., Senjyu, T., Sabory, N. R., Khosravy, M., Grilli, M. L., Mikhaylov, A., & Majidi, H. (2021). A forefront framework for sustainable aquaponics modeling and design. *Sustainability, 13*, 9313. https://doi.org/10.3390/su13169313

Dayong, N., Mikhaylov, A., Bratanovsky, S., Shaikh, Z. A., & Stepanova, D. (2020). Mathematical modeling of the technological processes of catering products production. *Journal of Food Process Engineering, 43*(2), 1–9.

Denisova, V., Mikhaylov, A., & Lopatin, E. (2019). Blockchain infrastructure and growth of global power consumption. *International Journal of Energy Economics and Policy, 9*(4), 22–29.

Dinçer, H., Yüksel, S., & Martinez, L. (2019). Balanced scorecard-based analysis about European Energy investment policies: A hybrid hesitant fuzzy decision-making approach with quality function deployment. *Expert Systems with Applications, 115*, 152–171.

Dinçer, H., Yüksel, S., Çağlayan, Ç., & Uluer, G. S. (2020a). The contribution of nuclear energy investment on sustainable financial and economic development. *Journal of Financial Economics and Banking, 1*(1), 39–51.

Dinçer, H., Yüksel, S., Ubay, G. G., & Karakuş, H. (2020b). BSC-Based evaluation for the factors affecting the performance of wind energy companies. In *Strategic priorities in competitive environments* (pp. 1–15). Springer.

Dong, B., Ikonnikova, I., Rogulin, R., Sakulyeva, T., & Mikhaylov, A. (2021). Environmental-economic approach to optimization of transport communication in megacities. *Journal of Environmental Science and Health, Part A, 56*(6), 660–666. https://doi.org/10.1080/10934529.2021.1913928

Dooyum, U. D., Mikhaylov, A., & Varyash, I. (2020). Energy security concept in Russia and South Korea. *International Journal of Energy Economics and Policy, 10*(4), 102–107.

Dorofeev, M. L. (2020). Impact of monetary policy on the level of economic inequality in the United States. *MGIMO Review of International Relations, 13*(5), 97–114. https://doi.org/10.24833/2071-8160-2020-5-74-97-114

Du, L., Dinçer, H., Ersin, İ., & Yüksel, S. (2020). IT2 fuzzy-based multidimensional evaluation of coal energy for sustainable economic development. *Energies, 13*(10), 2453.

Fang, S., Zhou, P., Dinçer, H., & Yüksel, S. (2021). Assessment of safety management system on energy investment risk using house of quality based on hybrid stochastic interval-valued intuitionistic fuzzy decision-making approach. *Safety Science, 141*, 105333.

Girma, S., Görg, H., & Strobl, E. (2007). The effect of government grants on plant level productivity. *Economics Letters, 94*(3), 439–444.

Grilli, M. L., Valerini, D., Rizzo, A., Yilmaz, M., Song, C., Hu, G., Chierchia, R., Mikhaylov, A., & Rinaldi, A. (2021). A comparative study of the mechanical and tribological properties of thin Al2O3 coatings fabricated by atomic layer deposition and radiofrequency sputtering. *Physica Status Solidi (A), 218*(24), 1–25.

Gura, D., Mikhaylov, A., Glushkov, S., Zaikov, M., & Shaikh, Z. A. (2020). Model for estimating power dissipation along the interconnect length in single on-chip topology. *Evolutionary Intelligence*, s12065.

Haghi, E., Raahemifar, K., & Fowler, M. (2018). Investigating the effect of renewable energy incentives and hydrogen storage on advantages of stakeholders in a microgrid. *Energy Policy, 113*, 206–222.

He, L., Liu, R., Zhong, Z., Wang, D., & Xia, Y. (2019). Can green financial development promote renewable energy investment efficiency? A consideration of bank credit. *Renewable Energy, 143*, 974–984.

Hoegen, A., Steininger, D. M., & Veit, D. (2018). How do investors decide? An interdisciplinary review of decision-making in crowdfunding. *Electronic Markets, 28*(3), 339–365.

Hong, J., Hong, S., Wang, L., Xu, Y., & Zhao, D. (2015). Government grants, private R&D funding and innovation efficiency in transition economy. *Technology Analysis & Strategic Management, 27*(9), 1068–1096.

Huang, Y., Masrur, H., Shigenobu, R., Hemeida, A. M., Mikhaylov, A., & Senjyu, T. (2021a). A comparative design of a campus microgrid considering a multi-scenario and multi-objective approach. *Energies, 14*, 2853. https://doi.org/10.3390/en14112853

Huang, Y., Yona, A., Takahashi, H., Hemeida, A. M., Mandal, P., Mikhaylov, A., Senjyu, T., & Lotfy, M. E. (2021b). Energy management system optimization of drug store operate electric vehicles charging station. *Sustainability, 14*, 1217836.

Ivanyuk, V., & Berzin, D. (2020). An algorithm for constructing an efficient investment portfolio. In R. Silhavy, P. Silhavy, & Z. Prokopova (Eds.), *Software engineering perspectives in intelligent systems. CoMeSySo 2020. Advances in Intelligent Systems and Computing* (Vol. 1294). Springer. https://doi.org/10.1007/978-3-030-63322-6_39

Ivanyuk, V., & Levchenko, K. (2020). Intelligent methods for predicting financial time series. In R. Silhavy, P. Silhavy, & Z. Prokopova (Eds.), *Software engineering perspectives in intelligent systems. CoMeSySo 2020. Advances in Intelligent Systems and Computing* (Vol. 1294). Springer. https://doi.org/10.1007/978-3-030-63322-6_41

Ivanyuk, V., Sunchalin, A., & Sunchalina, A. (2020). Development of an intelligent ensemble forecasting system. In R. Silhavy, P. Silhavy, & Z. Prokopova (Eds.), *Software engineering perspectives in intelligent systems. CoMeSySo 2020. Advances in Intelligent Systems and Computing* (Vol. 1294). Springer. https://doi.org/10.1007/978-3-030-63322-6_40

Ivanyuk, V. (2018). Econometric forecasting models based on forecast combination methods. In *Proceedings of 2018 11th International Conference Management of Large-Scale System Development, MLSD 2018*. https://doi.org/10.1109/MLSD.2018.8551825

Ivanyuk, V., & Soloviev, V. (2019). Efficiency of neural networks in forecasting problems. In *Proceedings of 2019 12th International Conference Management of Large-Scale System Development; MLSD 2019*. https://doi.org/10.1109/MLSD.2019.8911046

Koengkan, M., Fuinhas, J. A., & Vieira, I. (2020). Effects of financial openness on renewable energy investments expansion in Latin American countries. *Journal of Sustainable Finance & Investment, 10*(1), 65–82.

Lam, P. T., & Law, A. O. (2016). Crowdfunding for renewable and sustainable energy projects: An exploratory case study approach. *Renewable and Sustainable Energy Reviews, 60*, 11–20.

Lam, P. T., & Law, A. O. (2018). Financing for renewable energy projects: A decision guide by developmental stages with case studies. *Renewable and Sustainable Energy Reviews, 90*, 937–944.

Li, X., Zhu, S., Yüksel, S., Dinçer, H., & Ubay, G. G. (2020). Kano-based mapping of innovation strategies for renewable energy alternatives using hybrid interval type-2 fuzzy decision-making approach. *Energy, 211*, 118679.

Lisin, A. (2020). Valuation of the activities of foreign banks in the Russian banking sector. *Orbis, 15*(45), 53–63. http://www.revistaorbis.org.ve/pdf/45/art5.pdf

Liu, Z.-J., Panfilova, E., Mikhaylov, A., & Kurilova, A. (2021). COVID-19 crisis impact on the stability between parties in crowdfunding and crowdsourcing. *Wireless Personal Communications, 8932*. https://doi.org/10.1007/s11277-021-08932-z

Liu, Z.-J., Panfilova, E., Mikhaylov, A., & Kurilova, A. (2022). Assessing stability in relationship between parties in crowdfunding and crowdsourcing projects during the COVID-19 crisis. *Journal of Global Information Management, 30*(2), 1.

Melnichuk, A. V., Melnichuk, Y. A., Bocharov, A. V., Stepanova, D. I., & Kurbanov, S. A. (2020). Constitutional foundations of the functioning of local self-government. *Revista Inclusiones., 7*, 337–346.

Meynkhard, A. (2020). Effect of Bitcoin Volatility on Altcoins Pricing. *Advances in Intelligent Systems and Computing, 1294*, 652–664. https://doi.org/10.1007/978-3-030-63322-6_55

Meynkhard, A. (2019). Fair market value of bitcoin: Halving effect. *Investment Management and Financial Innovations, 16*(4), 72–85. https://doi.org/10.21511/imfi.16(4).2019.07

Mikhaylov, A., Danish, M. S. S., & Senjyu, T. (2021a). New stage in evolution of cryptocurrency market: Analysis by hurst method. In H. Dincer & S. Yuksel (Eds.), *Strategic outlook in business and finance innovation: Multidimensional policies for emerging economies* (pp. 35–45). Emerald Publishing Limited. https://doi.org/10.1108/978-1-80043-444-820211004

Mikhaylov, A. (2018a). Volatility spillover effect between stock and exchange rate in oil exporting countries. *International Journal of Energy Economics and Policy, 8*(3), 321–326.

Mikhaylov, A. (2018b). Pricing in oil market and using probit model for analysis of stock market effects. *International Journal of Energy Economics and Policy, 8*(2), 69–73.

Mikhaylov, A. (2018c). The yield of the carry trade strategy. *Finance: Theory and Practice, 22*(3), 52–63. https://doi.org/10.26794/2587-5671-2018-22-3-52-63

Mikhaylov, A. (2019). Oil and gas budget revenues in Russia after crisis in 2015. *International Journal of Energy Economics and Policy, 9*(2), 375–380.

Mikhaylov, A. (2020a). Cryptocurrency market analysis from the open innovation perspective. *Journal of Open Innovation: Technology, Market, and Complexity, 6*(4), 197. https://doi.org/10.3390/joitmc6040197

Mikhaylov, A. (2020b). Geothermal energy development in Iceland. *International Journal of Energy Economics and Policy, 10*(4), 31–35.

Mikhaylov, A. (2020c). Cryptocurrency market development: Hurst method. *Finance: Theory and Practice, 24*(3), 81–91.

Mikhaylov, A. (2021a). Development of Friedrich von Hayek's theory of private money and economic implications for digital currencies. *Terra Economicus, 19*(1).

Mikhaylov, A. (2021b). Lichens as indicators of atmospheric pollution in urban ecosystems. *Israel Journal of Ecology & Evolution, 67*(1–2), 60–68. https://doi.org/10.1163/22244662-bja10016

Mikhaylov, A. (2022). Efficiency of renewable energy generation in Russia. *Annals of the Brazilian Academy of Sciences, 94*(3).

Mikhaylov, A., & Tarakanov, S. (2020). Development of Levenberg-Marquardt theoretical approach for electric network. *Journal of Physics: Conference Series, 1515*, 052006.

Mikhaylov, A., Burakov, D., & Didenko, V. (2019). Relationship between Oil Price and Macroeconomic Indicators in Russia. *Finance: Theory and Practice, 23*(2), 105–116. https://doi.org/10.26794/2587-5671-2019-23-2-105-116

Mikhaylov, A., Sokolinskaya, N., & Nyangarika, A. (2018). Optimal carry trade strategy based on currencies of energy and developed economies. *Journal of Reviews on Global Economics, 7*, 582–592.

Mikhaylov, A., Yumashev, A., & Kolpak, E. (2021b). Quality of life, anxiety and depressive disorders in patients with extrasystolic arrhythmia. *Archives of Medical Science, 17*(1). https://doi.org/10.5114/aoms.2020.101359

Moiseev, N., Mikhaylov, A., Varyash, I., & Saqib, A. (2020). Investigating the relation of GDP per capita and corruption index. *Entrepreneurship and Sustainability Issues, 8*(1), 780–794.

Moiseev, N., Sorokin, A., Zvezdina, N., Mikhaylov, A., Khomyakova, L., & Danish, M. S. S. (2021). Credit risk theoretical model on the base of DCC-GARCH in time-varying parameters framework. *Mathematics, 9*, 2423. https://doi.org/10.3390/math9192423

Morkovkin, D., Lopatkin, D., Sadriddinov, M., Shushunova, T., Gibadullin, A., & Golikova, O. (2020a). Assessment of innovation activity in the countries of the world. *E3S Web of Conferences, 157*, 04015. https://doi.org/10.1051/e3sconf/202015704015

Morkovkin, D. E., Gibadullin, A. A., Kolosova, E. V., Semkina, N. S., & Fasehzoda, I. S. (2020b). Modern transformation of the production base in the conditions of Industry 4.0: problems and prospects. *Journal of Physics*: Conference Series, *1515*, 032014.

Mukhametov, A., Bekhorashvili, A., Avdeenko, A., & Mikhaylov, A. (2021). The Impact of Growing Legume Plants under Conditions of Biologization and Soil Cultivation on Chernozem Fertility and Productivity of Rotation Crops. *Legume Research, 44*, 12.

Mutalimov, V., Kovaleva, I., Mikhaylov, A., & Stepanova, D. (2021). Assessing regional growth of small business in Russia. *Entrepreneurial Business and Economics Review, 9*(3), 119–133. https://doi.org/10.15678/EBER.2021.090308

Nie, D., Panfilova, E., Samusenkov, V., & Mikhaylov, A. (2020). E-Learning financing models in Russia for sustainable development. *Sustainability, 12*(11), 4412.

Nyangarika, A., Mikhaylov, A., & Richter, U. (2019a). Influence oil price towards economic indicators in Russia. *International Journal of Energy Economics and Policy, 9*(1), 123–129.

Nyangarika, A., Mikhaylov, A., & Richter, U. (2019b). Oil price factors: Forecasting on the base of modified auto-regressive integrated moving average model. *International Journal of Energy Economics and Policy, 9*(1), 149–159.

Nyangarika, A., Mikhaylov, A., & Tang, B.-J. (2018). Correlation of oil prices and gross domestic product in oil producing countries. *International Journal of Energy Economics and Policy, 8*(5), 42–48.

Qiu, D., Dinçer, H., Yüksel, S., & Ubay, G. G. (2020). Multi-faceted analysis of systematic risk-based wind energy investment decisions in E7 economies using modified hybrid modeling with IT2 fuzzy sets. *Energies, 13*(6), 1423.

Radosteva, M., Soloviev, V., Ivanyuk, V., & Tsvirkun, A. (2018). Use of neural network models in market risk management. *Advances in Systems Science and Applications, 18*(2), 53–58. https://doi.org/10.25728/assa.2018.18.2.582

Ranjbar, O., Chang, T., Nel, E., & Gupta, R. (2017). Energy consumption and economic growth nexus in South Africa: Asymmetric frequency domain approach. *Energy Sources, Part B: Economics, Planning, and Policy, 12*(1), 24–31.

Rathnayaka, R. K. T., Seneviratna, D. M. K., & Long, W. (2018). The dynamic relationship between energy consumption and economic growth in China. *Energy Sources, Part B: Economics, Planning, and Policy, 13*(5), 264–268.

Saqib, A., Chan, T.-H., Mikhaylov, A., & Lean, H. H. (2021). Are the responses of sectoral energy imports asymmetric to exchange rate volatilities in Pakistan? Evidence from recent foreign exchange regime. *Frontiers in Energy Research, 9*, 614463. https://doi.org/10.3389/fenrg.2021.614463

Sunchalin, A. M., Kochkarov, R. A., Levchenko, K. G., Kochkarov, A. A., & Ivanyuk, V. A. (2019). Methods of risk management in portfolio theory. *Espacios, 40*(16), 8.

Uandykova, M., Lisin, A., Stepanova, D., Baitenova, L., Mutaliyeva, L., Yuksel, S., & Dincer, H. (2020). The social and legislative principles of counteracting ransomware crime. *Entrepreneurship and Sustainability Issues, 8*(2), 777–798.

Udalov, I. (2021). The transition to renewable energy sources as a threat to resource economies. *International Journal of Energy Economics and Policy, 11*(3), 1–6.

Uyeh, D. D., Asem-Hiablie, S., Park, T., Kim, K.-M., Mikhaylov, A., Woo, S., & Ha, Y. (2021). Could Japonica rice be an alternative variety for increased global food security and climate change mitigation? *Food, 10*, 1869. https://doi.org/10.3390/foods10081869

Varyash, I., Mikhaylov, A., Moiseev, N., & Aleshin, K. (2020). Triple bottom line and corporate social responsibility performance indicators for Russian companies. *Entrepreneurship and Sustainability Issues, 8*(1), 313–331.

Wang, S., Liu, Q., Yuksel, S., & Dincer, H. (2019). Hesitant linguistic term sets-based hybrid analysis for renewable energy investments. *IEEE Access, 7*, 114223–114235.

Yuan, G., Xie, F., Dinçer, H., & Yüksel, S. (2021). The theory of inventive problem solving (TRIZ)-based strategic mapping of green nuclear energy investments with spherical fuzzy group decision-making approach. *International Journal of Energy Research.*

Yüksel, S., Mikhaylov, A., & Ubay, G. G. (2021a). Factors causing delay in the installation of nuclear power plants. In S. Yüksel & H. Dinçer (Eds.), *Strategic approaches to energy management. Contributions to management science.* Springer. https://doi.org/10.1007/978-3-030-76783-9_7

Yüksel, S., Dinçer, H., Çağlayan, Ç., & Uluer, G. S. (2021b). Determining the optimal financial strategies for nuclear energy companies. In *Financial strategies in competitive markets* (pp. 1–16). Springer.

Yüksel, S., Mikhaylov, A., Ubay, G. G., & Uyeh, D. D. (2021c). The role of hydrogen in the black sea for the future energy supply security of Turkey. In S. Yüksel & H. Dinçer (Eds.), *Handbook of research on strategic management for current energy investments (pp. 1–15).* IGI Global. https://doi.org/10.4018/978-1-7998-8335-7.ch001

Yumashev, A., & Mikhaylov, A. (2020). Development of polymer film coatings with high adhesion to steel alloys and high wear resistance. *Polymer Composites, 41*(7), 2875–2880.

Yumashev, A., Ślusarczyk, B., Kondrashev, S., & Mikhaylov, A. (2020). Global indicators of sustainable development: Evaluation of the influence of the human development index on consumption and quality of energy. *Energies, 13*, 2768.

Zhao, Q., Cherkasov, S., Avdeenko, A., Kondratenko, L., & Mikhaylov, A. (2021). Integral estimate of the added value of agricultural crops in the synergetic agriculture on the example of vicia faba (Vicia faba L.). *Polish Journal of Environmental Studies, 30*(5), 1–9. https://doi.org/10.15244/pjoes/131949

Zhou, P., Luo, J., Cheng, F., Yüksel, S., & Dinçer, H. (2021). Analysis of risk priorities for renewable energy investment projects using a hybrid IT2 hesitant fuzzy decision-making approach with alpha cuts. *Energy, 224*, 120184.

Roles of FDI, Energy and Carbon Emission in Convergence or Divergence of Income in BRICS Nations in Neoclassical Growth Framework

Ramesh Chandra Das and Aloka Nayak

1 Introduction

Free trade through liberalizations and globalizations of the economies following the WTO accords has made tremendous growths of outputs and incomes of the nations of the so-called developed nations in the post-World War II era. The countries from the backward zones have started following the rising growth paths in the late 1980s and in the early 1990s of the last century. There had been increasing income disparities across the globe between the group of the developed economies of the west and the less developed nations from the east in the 1980s due to the endogenous growth factors such as knowledge capital generation, institutional factors, making the technological progress endogenous in place of the exogenous growth structure in the neoclassical model. The cross-country income differential among different countries at the world level has been going down, although there are rising disparities and inequalities in the individual country levels as well as across some groups of the economies (World Bank).

There is other side of the coin regarding the control and management of global economic and political powers. The so-called developed economies have occupied the authorities in different policy framing bodies at the global level, which according to the so-called less developed economies, has raised conflicts with the less developed or emerging economies. The global funds from the monetary authorities such as World Bank, IMF, etc. and the economic and financial ties among the developed nations have compelled some of the less developed countries to not making out of the poverty and unemployment traps. As solutions to this conflict and to fight against

R. C. Das (✉)
Vidyasagar University, Midnapore, West Bengal, India

A. Nayak
Department of Economics, Vidyasagar University, Midnapore, West Bengal, India

© The Author(s), under exclusive license to Springer Nature Switzerland AG 2022

H. Dinçer, S. Yüksel (eds.), *Sustainability in Energy Business and Finance*, Contributions to Finance and Accounting,
https://doi.org/10.1007/978-3-030-94051-5_15

the so-called western powers, there have been several economic groups that emerged from the eastern economic zones. Such a sound organization is BRICS, the combination of Brazil, Russia, India, China, and South Africa, whose attempt was to make a parallel consolidation for self defense against any external economic and political shocks. The acronym "BRIC" was initially formulated in 2001 by economist Jim O'Neill, of Goldman Sachs, in a report on growth prospects for the economies of Brazil, Russia, India, and China—which together represented a significant share of the world's production and population. South Africa was invited to join BRIC in December 2010, after which the group adopted the acronym BRICS. Together, BRICS accounts for about 40% of the world's population and about 30% of the GDP.

But it is now a natural question on whether the members of the BRICS are now in a position to raise their GDP levels as well as been capable of reducing the cross-country income differences within the group members. The present study has attempted to investigate the roles of the endogenous growth factors from economic and environmental fronts to justify their convergence or divergence in incomes of aggregate and per capita terms.

2 Review of Relevant Literature

The study reviews a list of selected works on the associations among the three indicators, FDI, energy use and CO_2 emissions, with GDP of the BRICS members. First studies on FDI-GDP, then Energy-GDP, and then CO_2-GDP are addressed, and at the end, the studies on income convergence among the BRICS nations are addressed.

Omri et al. (2014) investigate the causality link between CO_2 emissions, FDI and economic growth for 54 countries over the period 1990–2011 and found bidirectional causality between FDI and economic growth that an increase in the stock of FDI helps to promote economic growth and economic growth creates favorable conditions to attract FDI inflows into the considered regions. Also indicates a unidirectional positive relation from economic growth to CO_2 emissions that high economic growth leads to damage to the environmental quality. Zakarya et al. (2015) analyze how the emission of CO_2 in the BRICS countries affected by the variable factors, namely the total energy consumption, FDI, economic growth, and by using the co-integration tests and Granger causality test for the period 1990–2012, found that for the long-run energy consumption, GDP and FDI inflow increases CO_2 emissions. Though FDI inflow has no direct effect on CO_2 emission, but has a direct effect on GDP. Yaşar and Telatar (2018) analyze the relationship between FDI inflow and CO_2 emissions for 139 countries for the period 1970–2015. By applying Panel ARDL method and Granger causality test, they found that there is no significant short-run causality from FDI to CO_2 emissions for high income group.

Belke et al. (2010) try to examine the long-run relation between energy use and real GDP by applying Panel co-integration test and the Granger causality test for

25 OECD countries over the period 1981–2007, and the results assume bidirectional causal relationship between energy consumption and real GDP. Ouedraogo (2013) analyses the long-run relationship between economic growth and energy consumption for 15 African countries from 1980 to 2008 by using panel co-integration technique and found long-run and short-run unidirectional causality, where causality in short-run running from GDP to energy consumption and in long-run from energy consumption to GDP. Esen and Bayrak (2017) examine the effects of energy consumption on economic growth by means of a panel data analysis of 75 net energy-importing countries for the period 1990 to 2012, results indicate that there is a positive and statistically significant relationship between energy consumption and economic growth over the long-term such that energy consumption contributes more to economic growth as the import dependence of the country decreases.

Pao et al. (2011) examine the dynamic relationship between CO_2 emissions, energy use and real output for 1990–2007 for Russia by applying co-integration and causality test. The results from their Granger causality tests indicate the existence of strong bidirectional causality between these variables. The study also shows that energy use has a positive significant effect on CO_2 emissions, and real output has a negative impact on CO_2 emissions. Zhang and Zhou (2016) found that FDI inflow leads to reduce CO_2 emission in China over the period1995–2010 in 29 regions and suggest that impact of FDI on CO_2 emission varies by region. Azevedo et al. (2018) do a quantitative analysis between CO_2 emissions and GDP growth for the BRICS countries over the period 1980–2011 and find out that economic growth is the main driving force for the growth of CO_2 emissions in Brazil and Russia, and no significant relation appears between economic growth and CO_2 emissions for China, India, and South Africa.

For the income convergence analysis, Phiri (2018) worked on the per capita GDP convergence in BRICS nations using the time series technique for the period 1971–2015, and the results confirm on Brazil and China being the only two BRICS economies which present the most convincing evidence of per capita GDP converging back to its natural equilibrium after an economic shock, at the same time as Russia and South Africa provide less convincing evidence of convergence dynamics in the time series, and India having the weakest convergence features. The study by Basel and Rao (2018) examines the existence of absolute β convergence and σ convergence of real per capita GDP among the BRICS nations for 1990–2015 using static panel data model, namely the Fixed Effects and Random Effects for β convergence and coefficient of variation (CV) for sigma convergence. The results show the convergence among BRICS nations during the study period. But the results are not acceptable when we take quarterly data or extended GDP data. Applying the neoclassical growth and panel unit roots models on the quarterly data from 2006Q1–2017Q2 to 2009Q1–2017Q2, the study of Das, Das et al. (2019) reveals that there is no significant catching up of the countries in both the pre and post-BRICS periods, but there is conditional convergence in the first period through net FDI inflow and crude oil production. In another study, Das (2019) revisited the same issue for the same group and revealed that the countries are not unconditionally β converging but converging in conditional terms with the variables such as foreign direct investment (FDI) flow and working population. Furthermore, the study shows that the countries are converging in σ definition meaning the cross-country dispersion in per capita gross domestic product has fallen significantly. Hence, the formation of BRICS has made the countries relatively better off compared to pre-BRICS phase so far as the σ definition of convergence is concerned.

3 Analysis Results

The review of the literature so far does not exhibit the roles of the three crucial factors behind the remarkable growths of the member countries, FDI inflow, energy use and CO_2 emission, upon GDP and its convergence. The present study works on the issue of convergence of income in the group through these three crucial factors.

The present study uses the data of GDP (in current USD), Energy Use (in kg equivalent), Inflow of FDI (Foreign Direct Investment) (in current USD), CO_2 emission (in kt) and population taken from the open data source of the World Bank (www.worldbank.org). The data on population is used to derive the current value of per capita GDP (PCGDP) (in USD). The period of study is 1991–2020, and the member countries in the BRICS group are Brazil, Russia, India, China, and South Africa.

The study first presents the descriptive statistics on the average values of all the indicators across the pre-BRICS phase (1991–2009) and post-BRICS phase (2010–2020) and makes mean difference test through the student's t test.

For examining the cross-country convergence the study first reworks on the existence of non-convergence in aggregate GDP as well as PCGDP in terms of the absolute convergence hypothesis of the neoclassical growth model. Then it goes for sigma convergence tests. Finally, the study goes for investigating the causes of non-convergence of the incomes in the member countries by means of endogenous growth factors such as FDI inflow, energy uses and CO_2 emission.

Lets us recall the essence of the neoclassical theory on cross-country convergence. It is evident from the theory of absolute convergence that an economy with a lower income at the starting point will have a faster rate of growth. This implies that the income growth rate and initial income are inversely related. This is the notion of Absolute β Convergence. The same logic is applied to the four key variables of the present study.

We can now derive the expression for β in the form of cross-section regression. Let us suppose that there are n numbers of regions or countries within a geographical boundary with yit being the aggregate or PCGDP of the ith country. Consider the following regression model.

$$\log\left(y_{it}\right) = \alpha + (1 - \beta)\log\left(y_{i,t-1}\right) + u_{it} \quad (1)$$

It can be rewritten as

$$\log\left(y_{it}/y_{i,t-1}\right) = \alpha - \beta \log\left(y_{i,t-1}\right) + u_{it} \tag{2}$$

where α and β are constants respectively for intercept and slope with $\beta > 0$ and uit is a regular disturbance term following properties of normal distribution. In this equation, a positive sign of β means absolute convergence. That is, growth rate of income [log (yit/yi,t-1)] is inversely related to the initial income (yi, t-1). It is also to note that β is nothing but the slope of the income growth function. Here β also plays the role of speed of convergence.

Since the rate or speed of convergence depends on the gap between the initial value and the steady state value of the variable so it can be determined by the reorientation of the growth equation (Eq. 6) in the following form:

$$1/N\left[\log\left(y_{it}/y_{i,t-1}\right)\right] = \alpha - \left[\left(1 - e^{-\lambda t}\right)/N\right].\log\left(y_{i,t-1}\right) + e_{it} \tag{3}$$

where λ stands for the speed of convergence and N stands for total period of time under observation.

The hypothesis of Absolute β Convergence works well when the group of economies is homogeneous in the key parameters like savings ratio, population growth rate, rate of depreciation, etc., as considered in the Solow (1956) model.

The methodology of absolute β convergence by means of cross-country regressions has been criticized by Friedman (1992) and Quah (1993). They point out that these regressions are liable to produce biased estimates of β convergence, instead, the simple trend in the coefficient of variation of income provides an unbiased estimate of β convergence, which is known as σ convergence. The following regression equation can present the concept of σ Convergence -

$$\log\left(CV\right) = a + bt + u_t \tag{4}$$

where a is intercept constant, b is the slope constant or growth rate of CV over time and u is the random disturbance term. If the sign of "b" is found to be negative and statistically significant then we can say that the trend of CV is downward and that there is convergence among the regions or countries and that σ convergence exist.

If there are heterogeneities among the member countries in different indicators then the absolute convergence hypothesis does not work. The empirical data on the growth experiences of the so-called developed countries in the 1980s show that the countries did not converge to a common steady state, rather they diverged from this steady state value and converged to their own steady states determined from their individual growth determining factors (Barro & Sala-i-Martin, 1992). In other way to say that the countries follow conditional convergence hypothesis. The conditional convergence hypothesis for the BRICS group can be presented by the following equation containing the three conditioning factors, FDI, energy use and CO_2 level.

$$\log\left(Y_{it}/Y_{i,t\text{-}1}\right) = \alpha - \beta \log\left(Y_{i,t\text{-}1}\right) + \gamma \log\left(FDI_{i,t\text{-}1}\right) + \delta \log\left(\text{Energy}_{i,t\text{-}1}\right) + \theta \log\left(\text{CO}_{2_{i,t\text{-}1}}\right) + u_{it} \tag{5}$$

If, β is still positive and significant along with the significant values of γ, δ and θ then we can say that there is conditional convergence in individual country's incomes.

4 Discussion

At first, the study presents the key descriptive statistics (mean and standard deviation (SD)) of the four variables, PCGDP, per capita FDI inflow, per capita Energy use and per capita CO_2 emission, in both the pre-BRICs and post-BRICS phases. After that, it attempts to test the difference in the mean and SD from the pre- to post-BRICS phases to see the changes of the variables from non-alliance to alliance phase. And finally, the study attempts for the convergence analysis.

Table 1 presents the descriptive statistics and the test results for the mean and SD differences. It is observed that Russia becomes the leading country in average PCGDP followed by Brazil in both the phases. India stays at the trough in the list. Further, the rate of fluctuations in the PCGDP as measured by the SD shows that Russia's PCGDP is more fluctuating followed by Brazil in both the phases, whereas, India's fluctuation is the lowest in the two phases.

The results of mean difference in PCGDP show that all countries' PCGDP have significantly increased in the post-BRICS phase. But, there has been significant increase in the fluctuations of PCGDP in China only in the post-BRICS phase.

In terms of per capita FDI inflow, Brazil leads the group in both the phases followed by Russia in both the periods. India remains in the trough. The mean different test results show that, except South Africa, all the remaining four countries in the group have experienced significant increase in per capita FDI inflow to the countries in the post-BRICS period. But, for India and South Africa, the variances in FDI flow have increased.

With respect to the per capita energy use, Russia and South Africa top the list sequentially, and India still is at the bottom place. But, there are again significant increases in the per capita energy uses in the post-BRICS phase for all. Also, the rate of fluctuation has also increased for all in the second phase.

Finally, with respect to the per capita CO_2 emission, Russia and South Africa top the list sequentially, and India still is at the bottom place. There are significant increases in the per capita CO_2 emissions in the post-BRICS phase for all except Russia. Also, the rate of fluctuation has also increased for all in the second phase except Brazil and India.

Hence the series of the results on the descriptive statistical analysis show some degrees of associations of PCGDP with all the remaining three income influencing factors. Correlation analysis can provide better results to understand the degrees of

Table 1 Mean and Standard Deviation, and their statistical differences over the phases

Country	Pre-BRICS		Post-BRICS		t-value	F-test
	$Mean_1$	SD_1	$Mean_2$	SD_2	$Mean_{12}$	SD_{12}
	PCGDP					
Brazil	4659.16	1877.54	10328.48	2040.70	7.72(0.00)	1.18(0.36)
China	1324.14	1026.66	7909.34	1928.52	10.50(0.00)	3.53(0.00)
India	552.77	256.23	1690.93	263.98	11.60(0.00)	1.06(0.43)
Russian Federation	4119.20	2871.77	12123.02	2463.83	8.06(0.00)	0.74(0.31)
South Africa	4006.39	1171.39	6427.93	921.93	6.26(0.00)	0.62(0.22)
	Per Capita FDI Inflows					
Brazil	106.13	76.42	385.96	65.53	10.59(0.00)	0.74(0.32)
China	50.77	35.81	167.25	33.69	8.92(0.00)	0.89(0.44)
India	7.70	10.42	29.17	6.10	6.22(0.00)	0.34(0.04)
Russian Federation	99.95	149.53	241.00	132.62	2.68(0.01)	0.79(0.36)
South Africa	56.76	64.87	79.25	35.88	1.06(0.30)	0.31(0.03)
	Per Capita Energy Uses					
Brazil	1086.15	104.38	1461.14	54.89	11.01 (0.00)	0.28 (0.02)
China	1086.92	343.45	2187.24	90.77	10.35 (0.00)	0.07 (0.00)
India	421.49	50.31	618.35	27.84	11.91 (0.00)	0.31 (0.03)
Russian Federation	4552.31	466.83	4975.28	91.56	2.95 (0.01)	0.04 (0.00)
South Africa	2551.51	171.91	2691.08	41.94	2.63 (0.01)	0.06 (0.00)
	Per Capita CO_2 Emissions					
Brazil	0.0017	0.0002	0.0022	0.0002	7.84 (0.00)	0.84(0.40)
China	0.0033	0.0012	0.0071	0.0003	9.91(0.00)	0.06(0.00)
India	0.0009	0.0002	0.0016	0.0002	12.00(0.00)	0.79(0.36)
Russian Federation	0.0109	0.0012	0.0111	0.0003	0.53(0.60)	0.07(0.00)
South Africa	0.0071	0.0008	0.0078	0.0003	2.85(0.01)	0.19(0.01)

Source: Authors' calculations

Table 2 Correlation coefficients in the pairs of the variables in 1991–2020

		GDP	FDI	Energy	CO_2
Brazil	GDP	1	0.94	0.86	0.88
	FDI	0.94	1	0.90	0.91
	Energy	0.86	0.90	1	0.96
	CO_2	0.88	0.91	0.96	1
		GDP	FDI	Energy	CO_2
China	GDP	1	0.82	0.94	0.94
	FDI	0.82	1	0.93	0.94
	Energy	0.94	0.93	1	0.99
	CO_2	0.94	0.94	0.99	1
		GDP	FDI	Energy	CO_2
India	GDP	1	0.93	0.98	0.99
	FDI	0.93	1	0.94	0.93
	Energy	0.98	0.94	1	0.99
	CO_2	0.99	0.93	0.99	1
		GDP	FDI	Energy	CO_2
Russia	GDP	1	0.80	0.43	0.07*
	FDI	0.80	1	0.21*	−0.02*
	Energy	0.43	0.21*	1	0.89
	CO_2	0.07*	−0.02*	0.89	1
		GDP	FDI	Energy	CO_2
South Africa	GDP	1	0.54	0.89	0.91
	FDI	0.54	1	0.59	0.62
	Energy	0.89	0.59	1	0.96
	CO_2	0.91	0.62	0.96	1

Notes: * marks show insignificant results, and all the non * marks stand for significant results

associations. Table 2 gives the correlation coefficients of the different pairs of the variables.

The correlation coefficients as presented in the concerned table show that all the six pairs of the four variables are highly and positively correlated for all the four except Russia for the entire period of the study. The computed t values and their associated probability values for testing the significance of correlation are not shown in the table to avoid clumsiness. Hence, the four variables are related in BRICS context.

However, the correlation results for Russia show positive but insignificant results in the pairs of GDP-CO_2 and FDI-Energy. Also, the correlation is negative and insignificant for the pair FDI-CO_2. The insignificant results rule out the effect of the factors upon the PCGDP for the country.

The study revisits the convergence analysis for the elongated data sets for the PCGDP in the BRICS nations than the observations made by the studies such as Phiri (2018), Basel and Rao (2018), Das et al. (2019), Das et al. (2019), etc.

Estimating the sets of Eqs. (1, 2 & 3) the study obtains the results for convergence/divergence in absolute terms in PCGDP. The estimated equation is given below.

$$\text{AvgGrthPCGDP} = 20.88 - 2.02 \log\ \text{PCGDP}(1991)\ R\ \text{square}\ 0.69$$
$$\text{Probability}(0.09)(0.18)$$

The results show the signs of convergence as the coefficient of the regression coefficient (i.e., β) is negative, and the 69% of the variations in the average growth rates of per capita incomes of the member countries is explained by the past values (i.e., corresponding to the year 1991) of the per capita income. But the irony is that the derived value of the t statistics for the estimated β is 0.18, which allows to accept the null hypotheses of $\beta = 0$. This means the countries are not converging in the per capita income for the period 1991–2020. The countries are not catching up with the relatively stronger countries in the group.

As mentioned before, σ convergence captures the overall degree of disparity among the countries, and if such disparity goes down significantly over the year, then it is said that the countries are converging. Estimating Eq. 4 the results are given below.

$$\log CV = 4.42 - 0.017\, t\ R\ \text{square}\ 0.85$$
$$\text{Probability}(0.00)(0.00)$$

This means there is falling income disparity leading to sigma convergence but there is no catching up process as there is no absolute convergence. That is why it is required to investigate the other related variables which supported the falling dispersion property in terms of conditional convergence. These conditional factors are considered to be the FDI inflow, energy use and CO_2 emission. Following are the derived results for conditional convergence with respect to aggregate GDP and per capita GDP.

The study tests for the conditional convergence in both the aggregate GDP and per capita GDP to get the results from the broader perspectives. The results for the GDP show that the coefficient of β is negative and statistically significant along with the significant results of all the three conditional factors, FDI inflow, energy use and CO_2 emission. The signs of FDI and Energy are expectedly positive, meaning more FDI and more energy uses lead to more GDP, but the coefficient of CO_2 is negative, which means less CO_2 is associated with more GDP. The result for CO_2 is good so far as the goal of sustainable developmental goal is concerned.

$$\text{AvgGrthGDP} = -381.27 - 23.02 \log\ \text{GDP}(1991) + 4.31 \log\ \text{FDI} + 57.56$$

logEnergy

$$- 45.53 \log CO_2 \text{Probability}(0.00)(0.00)(0.00)(0.00)(0.00)\ R\ \text{square}\ 0.95$$

The results of conditional convergence for the per capita GDP are given in the following estimated regression equation.

$$\text{AvgGrthPCGDP} = -141.81 - 7.7\log\ \text{PCGDP}(1991) + 2.38\log\ \text{PCFDI} + 17.22$$

logPCEnergy

$$- 13.24\text{logPCCO2Probability}(0.00)(0.00)(0.00)(0.00)(0.00)R\ \text{square}\ \mathit{0.94}$$

It is observed that the sign of the estimated coefficient of β is negative and statistically significant along with the significant results of all the three conditional factors, FDI inflow, energy use and CO_2 emission. The signs of the coefficients of all the conditional factors are similar to the results under GDP.

Hence, it can be concluded that the members of the BRICS group do not follow the catching up process, their cross-country income differentials are going down leading to the notion of better-off ness of the members during the post-BRICS phase. Further, although there is no catching up in GDP or PCGDP, the countries are converging to their own steady states in incomes through conditional factors such as FDI flows, energy uses and CO_2 emissions. Therefore, the neoclassical prediction in cross-country convergence in income in absolute terms among the BRICS countries does not work, rather, the countries converge in terms of the conditional factors accompanying the foreign capital and environmental factors. The role of the initial period is important only when the factors such as foreign capital and environmental capital are taken into account. Therefore, foreign capital and environmental capital are working behind the huge growth of these highly developing group in the world. There is welcoming news that foreign capital worked well for the countries, but there is non-appreciable news that nature is getting exploited to get more level and growth of income of the countries. It is thus recommended to replace the natural capital by renewable energy and conservation capital to restore the environmental quality and to achieve the sustainable developmental goals.

5 Conclusion

The study revisited the income convergence among the BRICS countries with an elongated data set incorporating conditional growth factors such as FDI flow, energy use and CO_2 emission for the period 1991–2020. The results still do not establish the existence of absolute convergence in aggregate as well as per capita GDP, although there are sigma convergence. Further, the results of the cross-country regression incorporating the three conditional factors establish the significant conditional convergence in two types of incomes where these three factors have worked significantly in favor of the convergence dynamics. The results of the conditional convergence are good so far as economies' growth is concerned but not so good as it hampers the nature's stability or carrying capacity.

References

Azevedo, V. G., Sartori, S., & Campos, L. M. S. (2018). CO_2 emissions: A quantitative analysis among the BRICS nations. *Renewable and Sustainable Energy Reviews, 81*(July), 107–115.

Barro, R. J., & Sala-i-Martin, X. (1992). Convergence. *Journal of Political Economy, 100*, 223–251.

Basel, S., & Rao, R. P. (2018). β-Convergence of real per capita GDP in BRICS economies. *Indian Journal of Economics and Development, 14*(2), 309.

Belke, A, Dreger, C, & Haan, F.D. (2010). Energy consumption and economic growth: New Insight into the Co-integration relationship. *RUHR Economic papers 190,(June, 2010).RWI–Leibniz-InstitutfürWirtschaftsforschung, Ruhr-University Bochum, TU Dortmund University, University of Duisburg-Essen.*

Das, R. C., Das, U., & Das, A. (2019). BRICS nations and income convergence: An insight from the quarterly data for 2006Q1–2017Q2. *Global Business Review*. https://doi.org/10.1177/0972150918822057

Das, R. C. (2019). Is there cross-country income convergence among the BRICS nations? An Examination. *Journal of Infrastructure Development, 11*(1–2).

Esen, Ö., & Bayrak, M. (2017). Does more energy consumption support economic growth in net energy-importing countries? *Journal of Economics, Finance and Administrative Science, 22*(42), 75–98. https://doi.org/10.1108/JEFAS-01-2017-0015

Friedman, M. J. (1992). Do old fallacies ever die? *Journal of Economic Literature, 30*, 2129–2132.

Omri, A., Nguyen, K. D., & Rault, C. (2014). Causal interactions between CO_2 emissions, FDI, and economic growth: Evidence from dynamic simultaneous-equation models. *Economic Modelling, 42*(6 August, 2014), 382–389. https://mpra.ub.uni-muenchen.de/82504/

Ouedraogo, N. S. (2013). Energy consumption and economic growth: Evidence from the economic community of West African States (ECOWAS). *Energy Economics, 36*, 637–647. https://doi.org/10.1016/j.eneco.2012.11.011

Pao, H. T., Yu, H. C., & Yang, Y. H. (2011). Modeling the CO_2 emissions, energy use, and economic growth in Russia. *Energy, 36*(8), 5094–5100. https://doi.org/10.1016/j.energy.2011.06.004

Phiri, A. (2018). Robust analysis of convergence in per capita GDP in BRICS Economies., MPRA Working Paper, https://mpra.ub.uni-muenchen.de/86936/1/MPRA_paper_86936.pdf

Quah, D. T. (1993). Galton's fallacy and tests of the convergence hypothesis. *The Scandinavian Journal of Economics, 95*, 427–443.

Solow, R. M. (1956). A contribution to the theory of economic growth. *Quarterly Journal of Economics, 70*, 65–94.

Yaşar, N., & Telatar, M. E. (2018). The relationship between foreign direct investment and CO_2 emissions across a panel of countries. *Energy Economy, Finance and Geostrategy, 157–169*. https://doi.org/10.1007/978-3-319-76867-0_8

Zhang, C., & Zhou, X. (2016). Does foreign direct investment lead to lower CO_2 emissions? Evidence from a Regional Analysis in China. *Renewable and Sustainable Energy Reviews, 58*, 943–951.

Zakarya, G. Y., Mostefa, B., Abbes, S. M., & Seghir, G. M. (2015). Factors affecting CO_2 emissions in the BRICS Countries: A panel data analysis. *Procedia Economics and Finance, 26*(May), 114–125.

Key Issues for the Improvements of Shallow Geothermal Investments

Serhat Yüksel, Hasan Dinçer, Alexey Mikhaylov, Zafer Adalı, and Serkan Eti

1 The General Information for The Geothermal Energy

The world has been experienced a massive change in several decades. Regardless of the developed level, all countries have been trying to improve their economic and social infrastructures. The developed countries have allocated huge funds for research and development expenditures, innovations, and high value-added industrial fields to protect their position and improve their standard of life. Emerging countries and developing countries have been pursuing catching up with the economic life experienced in developed countries. All in all, the countries have endeavored to increase their economic and development level. One of the cornerstones of economic growth and development strategies is energy. All social and economic activities in the modern world are directly or directly integrated with energy. The dominant energy sources are nonrenewable energy resources involving coal, natural gas, petroleum, and petroleum derivatives (Mikayilov et al., 2020; Liu et al., 2021; Du et al., 2020). Although the countries can implement the economic and development policies through used nonrenewable energy resources to achieve the short-run objectives, the usage and processing of the nonrenewable energy resources will be the primary responsibility for the catastrophic. Nonrenewable energy resources induce environmental degradation, including CO_2 emission, land degradation,

S. Yüksel (✉) · H. Dinçer · S. Eti
The School of Business, İstanbul Medipol University, İstanbul, Turkey
e-mail: serhatyuksel@medipol.edu.tr; hdincer@medipol.edu.tr; seti@medipol.edu.tr

A. Mikhaylov
Financial University under the Government of the Russian Federation, Moscow, Russia

Z. Adalı
Artvin Çoruh University, Artvin, Turkey
e-mail: zaferadali@artvin.edu.tr

© The Author(s), under exclusive license to Springer Nature Switzerland AG 2022

H. Dinçer, S. Yüksel (eds.), *Sustainability in Energy Business and Finance*, Contributions to Finance and Accounting,
https://doi.org/10.1007/978-3-030-94051-5_16

habitat destruction, and water pollution. Concerning the detrimental effects of the world's nonrenewable energy resources, the future generation will be dystopic worlds in which nature, clean air, and various animal species will be a story. Besides, the nonrenewable energy resources are linked to a predictable end; in other explanations, utilization runs out of time. Alternatives energy resources have been examined and developed to sustain and improve economic life and protect the world for future generations. Renewable energy resources have been defined as the hope to overcome environmental degradation and become alternative energy sources for growth and development strategies. The environmental forces involving sun, wind, and aquifer provide renewable energy sources and unlimited supply (Dinçer & Yüksel, 2019a, b; Wang et al., 2019; Yüksel et al., 2019; Dinçer et al., 2019). Compared to nonrenewable energy resources, renewable energy is accepted as environmentally friendly because of fewer greenhouse gases and pollutions. There are various types of renewable energy resources; biomass, solar, wind, and hydropower. Geothermal energy resource has been received growing attention among the renewable energy resources.

Geothermal energy is a specific energy source that relies on a particular geographic position based on the near tectonic plates. Geothermal energy uses thermal fluids collected in rocks and fluids in the core of the earth (Malafeh & Sharp, 2015). The thermal fluids are achieved through artificial methods involving Flash and Double Flash power plants and Dry Stream power plants. The utilization of geothermal energy is dependent on the temperature of geothermal resources (Chamorro et al., 2012). Low temperature ranging 20–70 is employed for industry, producing chemicals, heating, and cooling. Intermediate temperature denoted 70–150 and high-temperature presenting above 150 is utilized for electricity generation and heating system. Geothermal energy resources have superior advantages over the other types of renewable energy resources because the weather conditions are not significant factors affecting the effectiveness of the resources (Mertoglu et al., 2003). Geothermal energy is a very predictable and reliable source to meet various human needs: heating, cooling, and electricity. Moreover, geothermal energy is a cleaner energy resource because some environmental degradations CO_2, NOX, and SOx emissions are significantly acceptable. In spite of the superior sides of geothermal energy, there are some risks resulting from the usage of geothermal energy, and pre-conditions establish a meeting for the geothermal plants (Labriet et al., 2015).

Furthermore, geothermal energy is one of the most promising renewable energy resources in electricity generation, heating, cooling, and direct utilization. There are some drawbacks, risks, and disadvantages to geothermal energy. The primary weakness in deep geothermal energy is that the specific location based on the active faults, volcanic, and tectonic plants provide the geothermal energy because the geothermal energy relies on the thermal fluids stored and heated by the earth's hearts. These properties induce the geothermal energy not to be available everywhere. However, the temperature of the thermal fluids has an impact on the utilization of geothermal energy and intermediate, and high-temperature fluids are required to use for heating, cooling, and electricity generation. Therefore, the development and the utilization of geothermal energy involve many pre-processing steps, which create uncertainty and risk. Geothermal energy is beginning with surface reconnaissance and exploration drilling, which need 2 to 3 years. The knowledge of explorations, adequate and competent human resources, the needs of the pieces of equipment for explorations are classified as faced risk factors. The early steps are the most significant risk because there is uncertainty over the characteristics of the geothermal resources' temperature and flow capacity, which determine the ability to drill productive wells and the usage of the geothermal (Speer et al., 2014). After the pre-step, the establishment of power plants and the thermal fluids cause some risks. The thermal fluids contain various detrimental chemicals: carbon dioxide, hydrogen sulfide, methane, mercury, arsenic, argon, and nitrogen. The injection of the thermal fluids induces the change in the water, surface, and subsurface temperature. However, nearly geothermal energy resources are adjacent to national parks, fertile agriculture, and highly intense urbanization. As a result, geothermal energy contains many risks ranging from the subsurface, surface, air emissions, and the drilling and well activities that can cause seismicity. All risks involving seismicity, water contamination, air pollution, noise, flora, and fauna damage lead to the lower social acceptability of geothermal than other renewable energy projects (Shortall et al., 2015; Bustaffa et al., 2020; Soltani et al., 2021).

There is also a new development in the geothermal energy fields. The improved system is called shallow geothermal energy, and it is accepted as one of the most promising energy resources, especially in heating, cooling, and warm water. In contrast to conventional (deep) geothermal energy, shallow geothermal energy is not restricted to a specific area based on active fault systems and tectonic plates. The sources of shallow geothermal do not rely on thermal fluids. Shallow geothermal is based on the geographic law in which the temperature is constant and 13c regardless of the seasons. The stored heat is provided by using simple techniques for heating and cooling. According to this side, Shallow geothermal energy is applied everywhere below 15-20 m depth. Without depending on thermal fluids, shallow geothermal has numerous advantages against conventional geothermal energy, especially in heating and cooling (Johnston et al., 2011). The negligible required initial capital is one of the leading superiors. Without the geothermal fluids involve various detrimental chemicals, Carbon dioxide, hydrogen sulfide, methane, mercury, arsenic, argon, and nitrogen, and the aquifer, the harm of shallow geothermal energy is fewer than the deep geothermal energy (Narsilio & Aye, 2018).

Although shallow geothermal energy is achieved everywhere with simplicity, lost cost devices, and its effects on the environment are negligible, it has been rare to use renewable energy resources. However, shallow geothermal energy will play a vital role in reducing the environmental degradation caused by household energy consumption, such as heating and cooling. In addition to the importance of generating electricity from renewable energy resources, heating and cooling are also essential in sustainable environment and development. IEA (2015) underlines that approximately 20% percent of global energy consumption in the world is associated with residential energy consumption, and along with these figures, Gi et al. (2018) predict that the need for energy-providing heating and cooling will be three-time higher than the amount experienced in 2010.

2 Shallow Geothermal Systems

Shallow geothermal applications have been received massive attention in Europe in the last decades. The natural underground abilities generate the shallow geothermal for providing constant temperature, which induces the groundwaters' temperature to increase. In contrast to deep geothermal energy, the low enthalpy energy source is possible to utilize as heating and cooling needs of buildings. It is underlined that a notable benefaction to environmental health can be initiated with a decrease of fossil fuel consumption by revealing this "hidden treasure," pure until today, by means of shallow geothermal systems, and to the national economy with the reduction in the amount of energy imported and turning to our renewable own resources. There are around 1.9 million shallow geothermal systems recently in practice in Europe. Shallow geothermal energy composes 66.5% of geothermal use in Europe. Furthermore, the energy received in European countries with these systems has attained a total capacity of 26,900 MWth (megawatt thermal). The utilization of shallow geothermal energy in the world has been varied in terms of heating and cooling. In middle Europe, the shallow energy system is principally installed to the heating, but the application of shallow energy aiming to free cooling and active cooling has been employed in larger commercial installations. As for Northern Europe, shallow geothermal energy is preferred for heating, while shallow geothermal energy abilities for cooling have been received more attention than heating in Southern Europe. In Scandinavia, the commercial installations based on high cooling loads have been determined applications.

Shallow geothermal has a superior appearance compared to nonrenewable and other renewable energy resources. The prime advantage is that the below 15–20 m depth provides shallow geothermal energy. Though the effectiveness of the most renewable energy resources such as solar totally relies on climatic conditions such as sunny days, gale forces, beachcomber, shallow geothermal energy seem to be more independent of climatic conditions. As for the economic advantages, there is no requirement for substantial investment costs in shallow geothermal energy, and it has a low operating cost and longer life expectancy. Though shallow geothermal energy does not generate electricity like deep geothermal energy, it is used for heating, cooling, and warn water provision. According to Eurostat (2018), the approximately quarterly final energy consumption in the EU in 2016 is associated with household energy consumption. Nearly 80%of households' energy consumption is related to building heating and heating water. The same report also underlines that renewable energy sources provide only 16% of the household's energy. Within this objective, it is evident that strengthening the share of shallow geothermal energy for building heating, cooling, and warm water provision becomes a vital piece of the solutions for decarbonization (Sanner et al., 2003; Sanner, 1987).

Although geothermal energy is described as the energy achieving from the form of heat in the depths of the earth, shallow geothermal energy is defined as the system used heat pump in the heating and cooling, which allows the use of low-temperature groundwater in the shallow depths. Therefore, shallow geothermal energy is available everywhere, equipped with shallow aquifers providing energy sources with low enthalpy. The earth performs as a collector for the storage of energy achieving from the sun under the ground. Although Deep geothermal energy is totally derived from reservoirs heating by plate tectonics and the earth's core, shallow geothermal energy is mainly derived from sun rays, and only a tiny proportion of the stored energy in shallow geothermal energy is generated by internal heat or heat generated by plate tectonics. Shallow geothermal usage primarily covers the part of the ground until 100 m depth, but the system is available to apply to deeper areas; however, this raises the cost. The region of the climatic conditions is one of the most critical factors impacting the usage of shallow geothermal energy. Suppose a region's climatic condition is based on Hot and dry summers and quite rainy in winter. In that case, the aquifer in the area has essential potential in terms of groundwater resources. The grounds absorb the sun rays, and the calescent ground increases the groundwater temperature (Rybach, 2010; Sanner et al., 2003; Sanner, 1987).

Shallow geothermal systems are based on the working principle in which the stored ground heat from the ground surface down is harnessed by means of employing various methods. Shallow geothermal usage is not associated with specific geothermal anomalies. Regarding this property, it can be claimed that everywhere below the surface provides shallow geothermal energy resources. The underground offers a reserved and astonishingly large heat source. It is possible to use Heat storage and heat sink under the earth for the geothermal heat pump. Shallow geothermal energy relies on the geophysical law based on the underground abilities for heat sink and heat storage. The scientific evidence of the underground properties is proven by Antoine-Laurent de Lavoisier (1743–1794), who is the founder of modern chemistry. Lavoisier observed the ground at a depth of 27 m below street level with the help of installing a mercury thermometer. According to the experiment, it is documented that the temperature at the observed depth is consistent throughout the year. Later, a growing body of experiments in terms of the underground heat sink has been conducted by various researchers. For example, Alexander von Humboldt made an observation in Paris. As a result of the statement, it is noted that the average temperature equals 12 C, and the change in the temperature over the seasons only varies only 0.04 C. Although the underground geothermal properties are proven, the first time used heat was stored underground for heating occurred in the mid-twentieth century. The first plant installation was recorded in Indianapolis, the USA, in 1945. Since then, a tremendous ground source heat pump has been experienced in some European countries like Sweden, Switzerland, and Germany. Several mechanic studies and observations have been conducted to exploit the earth as the heat source and heat storage for heat pumps in the following years. However, an article back in 1947 demonstrates all techniques employing until today. According to this article, Groundwater wells, horizontal coils, vertical boreholes, U-pipe, and spiral forms are the leading techniques to exploit the heat source. In addition, two major methods are bringing shallow geothermal energy to good use. A heat pump is a device transferring heat from a lower temperature to a higher temperature, and it has one of the most effective methods in shallow geothermal energy. The heat pump is also called Ground Source

Heat Pumps, denoted GSHP. There are three primary components in the application of the GSHP. The first component is that the heat sink under the ground is revealed or eradicated in terms of usage of shallow geothermal for heating or cooling. The achieving shallow geothermal reservoir is converted into a suitable temperature level in the second component. The achieved and converted shallow geothermal sources are transferred to heat or cold rooms in the final system. Another method is Underground Thermal Energy Storage System, denoted as UTES. The storage of heat or cold artificially changes the stored temperature through the UTES system. On the other hand, the Heat pump system is the dominant figure in applying shallow geothermal energy (Eugster & Sanner, 2007; Sanner et al., 2003).

A heat-up machine allows heat transport from the inexpedient temperature to a suitable temperature in terms of heating or cooling. Within this purpose, a driving compressor is applied as external energy. The working principle of the heat pump is based on the thermodynamic postulate. The thermodynamic postulate is that the temperature of a gas increase when it is compressed into a smaller volume. In a heat pump, refrigerant is diffused by the heat sink, and the external energy typically achieved from electric power compresses the resulting gas. Therefore, the hot gas is transferred to the heating system. Then the used gas condenses again to a liquid, and the fluid comes back into the low-pressure area and becomes cold. The process continues the mentioned cycle. A ground source heat pump system is also used for cooling objects. Although the heating mode has been popular in Western Europe for many years, the reversible heat pump for cooling has been accepted as a new one in shallow geothermal usage. The cooling mode is called the Geoexchange system, and the first application was built in Germany in 1987. There are more required factors in the cooling purpose in contrast to the heating mode. For example, cooling is only achievable if the cooling load is more miniature than the heating load. Besides, the climate has become a matter of factors. Arid weather is wanted conditions; otherwise, the heat pump has to perform as a chiller, or supplementary de-humidification should be applied to cool (Eugster & Sanner, 2007; Öngen & Ergüler, 2021).

Heat pumps are the essential instruments in shallow geothermal which transfer heat from heat sources to a cooler. The heat pump is based on the working principle in which water enters and leaves and its temperature is increased/decreased. There are two ways of establishing a heat pump system; open-loop and closed-loop systems. The closed-loop system has been chiefly utilized as a shallow geothermal mechanism. The closed-loop mechanism considers the ground as the heat source. The soil temperature is obtained from the ground with the help of pipes horizontally or vertically installed under the ground. The fluid named the thermal transfer fluid consisting of water and antifreeze circulates continuously in the closed-loop system within this application. Heat pumps increase the heat obtained from the ground due to the heat exchange of the antifreeze fluid. In the closed-loop system, the pipes are laid horizontally to a depth of 1–2 meters. In contrast, vertically piper laid depends on the application's convenience and the purpose of heating or cooling and generally vertically up to a depth of 100 meters. The open-loop system has been less applied compared to the closed-loop system. The heat energy is achieved from the groundwater, rivers, and other water sources through pipes laid in the borehole. The directly achieved fluids are transferred to the heat exchanger and next to the heat pump. Water is pumped through the borehole, and water circulates between two or more groundwater wells. Drawed water from underground is used for heating or cooling, and used water injects back into the same aquifer through a second borehole. The aquifer's high permeability features play an important role in minimizing the drawdown when the required amount of water is extracted in the open-loop system (Hähnlein et al., 2013).

Choosing a suitable GSHP system depend on various factors. The features of geology and hydrogeology perform a vital role in the right installation system. If the geologic features of the underground have sufficient permeability, open systems are recommended to install for the utilization of shallow geothermal energy. Area and utilization on the surface, the heating and cooling aspects of the buildings, etc., are also important determinants of the specific installation system.

In the open system, freely flowing underground waters, rivers, and other types of water sources heating by the solid earth perform as heat sources. One water well is installed to extract water, and the other well is installed to re-inject it into the same aquifer. Though an open system is a powerful method to exploit powerful heat sources at comparably low cost; required some maintenances and the existence of a suitable aquifer are determinants limiting the open system. In addition to the sufficient permeability of the aquifers, the composing of the chemical components in the aquifers becomes a preventive factor. For example, a less harmful substance such as low iron content should be noticed before utilizing the open system to avoid problems with corrosion and clogging. Furthermore, an open system is unsuitable for smaller installations, and a more extensive installation is fitted with an open system. The most powerful open heat-up system supplied ca. 10 MW has been operated in Louisville, Kentucky, USA, to heat and cold hotels and offices (Eugster & Sanner, 2007).

Another method for the usage of shallow geothermal energy is a closed system. The closed systems can be installed horizontally or vertically. The vertical system is the oldest version of the closed system. The scientific evidence above mentioned the constant temperature below a certain depth is achieved through a vertically laid borehole system. Moreover, the land restriction and available area push the new technologies to improve. Horizontal methods in the closed system seem to have numerous advantages in terms of land usage. Especially, western and Central Europe, where the restriction in the area and the land are so expensive, prefer some unique horizontal methods to exploit the shallow geothermal for heating and cooling through relatively dense laid individual pipes (Eugster & Sanner, 2007; Sanner, 1999; Öngen & Ergüler, 2021).

Climate conditions play a vital role in the applications of renewable energy resources. For example, the experienced of sunny days is the primary determinant of solar energy, and the gale forces and the ways of the wind are the primary indicators of wind turbines. Although Shallow geothermal energy is less dependent on climate conditions, two primary climate conditions have power limited the applications of shallow geothermal. For example, the extreme humidity level in the summer times wherrets the cooling process achieving the ground to a building.

The average temperature in the air is also another essential condition because the sunny rays provide the earth's ambient temperature. Another concern over shallow geothermal energy is environmental limitations, though shallow geothermal energy induces less ecological degradation. The shallow geothermal energy less cause global emissions of carbon dioxide and other detrimental ecological substances. In contrast, some contamination can be experienced in the installations and operational times of the GSHP system. For example, drilling, boreholes laid can damage buildings, fauna, and flora. Heating and cooling operational are possible factors that change the underground temperature, and hence the ground chemistry and bacterial composition of the underground will be affected. The usage of antifreeze and the chemical composition of boreholes are other stimulation contaminated the surface, subsurface, and aquifers. Furthermore, the defined depth of the underground provides shallow geothermal energy; in other explanations, shallow geothermal is feasible everywhere. Nevertheless, what types of shallow geothermal energy applications require some geological conditions. Closed-loop systems seem to be generally appropriate in all kinds of ground, but thermal properties and drilling may be regarded as factors impacted by geology. Horizontal loops are placed at a near-surface varied depth between 1 and 1.5 m. Generally, the tube is made of Polyethylene, and its size is up to 25 mm. Within this context, the horizontal loop does not negatively impact the groundwater if this system is correctly dimensioned and built. Furthermore, the open system relies on more geological conditions because the existence of aquifer, hydraulic properties, and the chemical components of water chemistry perform limitation factors on opens-systems. The several perfectives of hydrogeological conditions have a direct impact on the weel capacity of the open system. Unfavorable chemical compositions in aquifers, the aquifer's size and geometry, and the current user area of the aquifers such as drinking water or irrigation are the critical parameters preventing open -system. Despite the preventing factors based on hydrogeological limitations, closed-loop systems are performed as an alternative (Eugster & Sanner, 2007; Öngen & Ergüler, 2021; Rybach & Eugster, 2010).

The growing current market activities in shallow geothermal energy and the risks of shallow geothermal energy concerning environmental degradation have caused the implementation of some legislation and the required license for GSHP installation (Hähnlein et al., 2013). the abovementioned ecological risk resulting from shallow geothermal energy is classified into two main principles named groundwater and subsurface contaminations (Rügner et al., 2006). Regarding the countries with a progressed GSHP market consisting of Germany, Sweden, Switzerland, and Austria, the authorities related to water management have published guidelines for the construction and operation of GSHP installations (Eugster & Sanner, 2007). Furthermore, the comprehensive guidelines and standards are also determined due to some researcher's findings (Sanner, 2008; Hähnlein et al., 2011). There is various license in the application ranging from the building permits, the mining, the water management, and protection. Moreover, the relevant authorities have provided the determined risk potential of the installation and the map used to determine the suitable GSHP methods. All legal frameworks, the Treaty on the Functioning of the European Union (TFEU, 2010) and the European Water Framework Directive (EU-WFD, 2000), are the most important technical and legal frameworks. If all information and instruction provided by the authorities, the location, and the indication of which type of GSHP application are followed by the market players, the risks of shallow geothermal energy can be mitigated.

3 Influencing Factors of Shallow Geothermal Energy Investments

In this study, it is aimed to determine the factors that are important for increasing shallow geothermal energy investments. In this framework, balanced scorecard-based factors (finance-C1, customer-C2, internal processes-C3, research & development-C4) were taken into account (Xie et al., 2021; Delen et al., 2020; Yuan et al., 2020; Zhou et al., 2020; Dinçer et al., 2020). The DEMATEL method was used to determine the importance weights of these factors. This method has been preferred in many different analyzes in the literature (Jun et al., 2021; Yuan et al., 2021; Haiyun et al., 2021; Fang et al., 2021; Ding et al., 2021; Dínçer et al., 2021). The details of the analysis results are given in Table 16.1.

Table 16.1 demonstrates that research and development is the most crucial balanced scorecard parameter for the improvement of shallow geothermal projects. Additionally, the internal process has the second-highest weight. On the other side, finance and customer have lower significance in comparison with others.

4 Conclusion

This chapter focuses on the ways to improve shallow geothermal investment projects. For this purpose, balance scorecard parameters are evaluated by considering DEMATEL methodology. It is identified that research and development is the most crucial balanced scorecard parameter for the improvement of shallow geothermal projects. Thus, it is strongly recommended that countries should give priority to the research and development works in this regard. This situation has a positive contribution to implement up-to-date technologies for these projects. Hence, high-cost problem of the geothermal energy investments can be minimized.

Table 16.1 Analysis Results

Factors	Weights
Finance-C1	0.2465
Customer-C2	0.2024
Internal Processes-C3	0.2572
Research & Development-C4	0.2829

References

Bustaffa, E., Cori, L., Manzella, A., Nuvolone, D., Minichilli, F., Bianchi, F., & Gorini, F. (2020). The health of communities living in proximity of geothermal plants generating heat and electricity: A review. *Science of the Total Environment, 706*, 135998.

Chamorro, C. R., Mondéjar, M. E., Ramos, R., Segovia, J. J., Martín, M. C., & Villamañán, M. A. (2012). World geothermal power production status: Energy, environmental and economic study of high enthalpy technologies. *Energy, 42*(1), 10–18.

Delen, D., Dorokhov, O., Dorokhova, L., Dinçer, H., & Yüksel, S. (2020). Balanced scorecard-based analysis of customer expectations for cosmetology services: a hybrid decision modeling approach. *Journal of Management Analytics, 7*(4), 532–563.

Dinçer, H., & Yüksel, S. (2019a). Multidimensional evaluation of global investments on the renewable energy with the integrated fuzzy decision-making model under the hesitancy. *International Journal of Energy Research, 43*(5), 1775–1784.

Dinçer, H., & Yüksel, S. (2019b). Balanced scorecard-based analysis of investment decisions for the renewable energy alternatives: A comparative analysis based on the hybrid fuzzy decision-making approach. *Energy, 175*, 1259–1270.

Dinçer, H., Yüksel, S., & Martinez, L. (2019). Balanced scorecard-based Analysis about European Energy Investment Policies: A hybrid hesitant fuzzy decision-making approach with Quality Function Deployment. *Expert Systems with Applications, 115*, 152–171.

Dínçer, H., Yüksel, S., & Martínez, L. (2021). House of quality-based analysis of new service development using context free grammar evaluation-enhanced fuzzy hybrid modelling. *IEEE Access*.

Dinçer, H., Yüksel, S., Ubay, G. G., & Karakuş, H. (2020). BSC-based evaluation for the factors affecting the performance of wind energy companies. In *Strategic priorities in competitive environments* (pp. 1–15). Springer.

Ding, Z., Yüksel, S., & Dinçer, H. (2021). An Integrated Pythagorean fuzzy soft computing approach to environmental management systems for sustainable energy pricing. *Energy Reports, 7*, 5575–5588.

Du, L., Dinçer, H., Ersin, İ., & Yüksel, S. (2020). IT2 fuzzy-based multidimensional evaluation of coal energy for sustainable economic development. *Energies, 13*(10), 2453.

Eugster, W. J., & Sanner, B. (2007, May). Technological status of shallow geothermal energy in Europe. *In Proceedings European Geothermal Congress, 8*.

EUWFD, 2000.European Water Framework Directive, 2000/60/EC. European Parliament.

Fang, S., Zhou, P., Dinçer, H., & Yüksel, S. (2021). Assessment of safety management system on energy investment risk using house of quality based on hybrid stochastic interval-valued intuitionistic fuzzy decision-making approach. *Safety Science, 141*, 105333.

Gi, K., Sano, F., Hayashi, A., Tomoda, T., & Akimoto, K. (2018). A global analysis of residential heating and cooling service demand and cost-effective energy consumption under different climate change scenarios up to 2050. *Mitigation and Adaptation Strategies for Global Change, 23*(1), 51–79.

Hähnlein, S., Bayer, P., Ferguson, G., & Blum, P. (2013). Sustainability and policy for the thermal use of shallow geothermal energy. *Energy Policy, 59*, 914–925.

Hähnlein, S., Blum, P., & Bayer, P. (2011). Shallow geothermal energy-current legal situation. *Grundwasser, 16*(2), 69–75.

Haiyun, C., Zhixiong, H., Yüksel, S., & Dinçer, H. (2021). Analysis of the innovation strategies for green supply chain management in the energy industry using the QFD-based hybrid interval valued intuitionistic fuzzy decision approach. *Renewable and Sustainable Energy Reviews, 143*, 110844.

IEA. (2015). *Energy Balances of OECD/Non-OECD Countries*.

Johnston, I. W., Narsilio, G. A., & Colls, S. (2011). Emerging geothermal energy technologies. *KSCE Journal of Civil Engineering, 15*(4), 643–653.

Jun, Q., Dinçer, H., & Yüksel, S. (2021). Stochastic hybrid decision-making based on interval type 2 fuzzy sets for measuring the innovation capacities of financial institutions. *International Journal of Finance & Economics, 26*(1), 573–593.

Labriet, M., Joshi, S. R., Vielle, M., Holden, P. B., Edwards, N. R., Kanudia, A. ... Babonneau, F. (2015). Worldwide impacts of climate change on energy for heating and cooling. *Mitigation and Adaptation Strategies for Global Change, 20*(7), 1111–1136.

Liu, Y., Gong, X., Yüksel, S., Dinçer, H., & Aydın, R. (2021). A multidimensional outlook to energy investments for the countries with continental shelf in East Mediterranean Region with Hybrid Decision Making Model based on IVIF logic. *Energy Reports, 7*, 158–173.

Malafeh, S., & Sharp, B. (2015). Role of royalties in sustainable geothermal energy development. *Energy Policy, 85*, 235–242.

Mertoglu, O., Bakir, N., & Kaya, T. (2003). Geothermal applications in Turkey. *Geothermics, 32*(4–6), 419–428.

Mikayilov, J. I., Mukhtarov, S., Dinçer, H., Yüksel, S., & Aydın, R. (2020). Elasticity analysis of fossil energy sources for sustainable economies: A case of gasoline consumption in Turkey. *Energies, 13*(3), 731.

Narsilio, G. A., & Aye, L. (2018). Shallow geothermal energy: An emerging technology. In *Low Carbon Energy Supply* (pp. 387–411). Springer.

Öngen, A. S., & Ergüler, Z. A. (2021). The effect of urban heat island on groundwater located in shallow aquifers of Kütahya city center and shallow geothermal energy potential of the region, Turkey. *Bulletin of the Mineral Research and Exploration, 165*(165), 217–234.

Rügner, H., Finkel, M., Kaschl, A., & Bittens, M. (2006). Application of monitored natural attenuation in contaminated land management–a review and recommended approach for Europe. *Environmental Science & Policy, 9*(6), 568–576.

Rybach, L. (2010, April). Status and prospects of geothermal energy. In *Proceedings World Geothermal Congress* (Vol. 5, pp. 1–5).

Rybach, L., & Eugster, W. J. (2010). Sustainability aspects of geothermal heat pump operation, with experience from Switzerland. *Geothermics, 39*(4), 365–369.

Sanner, B. (1987). *Research on ground-coupled systems with vertical heat exchangers*. Proc. WS on GSHP, Albany, Report IEA-HPC, 113–117.

Sanner, B. (1999). *High temperature underground thermal energy storage*. State-of-the-art and prospects.

Sanner, B. (2008, May). Guidelines, standards, certification and legal permits for ground source heat pumps in the European Union. In *9th International IEA Heat Pump Conference* (Vol. 9).

Sanner, B., Karytsas, C., Mendrinos, D., & Rybach, L. (2003). Current status of ground source heat pumps and underground thermal energy storage in Europe. *Geothermics, 32*(4–6), 579–588.

Shortall, R., Davidsdottir, B., & Axelsson, G. (2015). Geothermal energy for sustainable development: A review of sustainability impacts and assessment frameworks. *Renewable and Sustainable Energy Reviews, 44*, 391–406.

Soltani, M., Kashkooli, F. M., Souri, M., Rafiei, B., Jabarifar, M., Gharali, K., & Nathwani, J. S. (2021). Environmental, economic, and social impacts of geothermal energy systems. *Renewable and Sustainable Energy Reviews, 140*, 110750.

Speer, B., Economy, R., Lowder, T., Schwabe, P., & Regenthal, S. (2014). *Geothermal exploration policy mechanisms: Lessons for the United States from international applications* (No. NREL/TP-6A20–61477). National Renewable Energy Lab.(NREL), Golden, CO (United States).

TFEU. (2010). Consolidated Version of the Treaty on the Functioning of the *European Union, 2010/C83/01*. European parliament.

Wang, S., Li, W., Dincer, H., & Yuksel, S. (2019). Recognitive approach to the energy policies and investments in renewable energy resources via the fuzzy hybrid models. *Energies, 12*(23), 4536.

Xie, Y., Zhou, Y., Peng, Y., Dinçer, H., Yüksel, S., & an Xiang, P. (2021). An extended pythagorean fuzzy approach to group decision-making with incomplete preferences for analyzing balanced scorecard-based renewable energy investments. *IEEE Access, 9*, 43020–43035.

Yuan, G., Xie, F., Dinçer, H., & Yüksel, S. (2021). The theory of inventive problem solving (TRIZ)-based strategic mapping of green nuclear energy investments with spherical fuzzy group decision-making approach. *International Journal of Energy Research.*

Yuan, J., Zhang, Z. M., Yüksel, S., & Dinçer, H. (2020). Evaluating recognitive balanced scorecard-based quality improvement strategies of energy investments with the integrated hesitant 2-tuple interval-valued Pythagorean fuzzy decision-making approach to QFD. *IEEE Access, 8*, 171112–171128.

Yüksel, S., Dinçer, H., & Meral, Y. (2019). Financial analysis of international energy trade: a strategic outlook for EU-15. *Energies, 12*(3), 431.

Zhou, P., Zhou, P., Yüksel, S., Dinçer, H., & Uluer, G. S. (2020). Balanced scorecard-based evaluation of sustainable energy investment projects with it2 fuzzy hybrid decision making approach. *Energies, 13*(1), 82.

Religious Principles for the Development of Energy Investments

Nikita Makarichev, Tomonobu Senjyu, and Sergey Prosekov

1 Introduction

Ethical the values and philosophical ideas of the Buddhist tradition can undoubtedly be useful for effective business development. For this, some concepts of Buddhism need to be translated into a language accessible to western people, including them in the sphere of business communication. Both eastern and western are worthy of special attention to experience of using Buddhist philosophy and ethics in the formation of corporate culture and doing business.

Freedom of action, orientation on commercial success, and profit at any cost today tearing stable to still a system of capitalism.

In the conditions of modern crisis processes in the global and domestic economies the ethics of socioeconomic life as a whole again comes to the fore and business in particular. The importance of an ethical approach to the energy management is increasingly addressed not only by religious leaders, but also by well-known public figures and politicians. The understanding of the fact that crisis phenomena in the energy management are a consequence of the global crisis of modern civilization a nation that "forgot" about universal moral precepts and traditional spiritual values.

N. Makarichev
Financial Research Institute of the Ministry of Finance of the Russian Federation, Moscow, Russia

T. Senjyu
Department of Electrical and Electronics Engineering, University of the Ryukyus, Nishihara, Japan
e-mail: b985542@tec.u-ryukyu.ac.jp

S. Prosekov (✉)
Financial University under the Government of the Russian Federation, Russian Federation, Moscow, Russia

© The Author(s), under exclusive license to Springer Nature Switzerland AG 2022

H. Dinçer, S. Yüksel (eds.), *Sustainability in Energy Business and Finance*, Contributions to Finance and Accounting,
https://doi.org/10.1007/978-3-030-94051-5_17

Problems faced by modern technogenic civilization indicate the need for reassessment of values in the spiritual sphere. In this regard, an appeal to the old experience of world denominations, in particular to Buddhist culture.

This concept contributed to the understanding of the labor process as specific collective activity, where the collective is compared with individual individuals has a higher level soul.

The role of ethical standards of Buddhism in the formation of the Japanese doctrine of corporate social responsibility is analyzed. According to the author, a wide the spread of corporate social responsibility ideas in Japan was due to the fact that these principles were adopted by Japanese business circles as a continuation of their own ethical and religious traditions.

2 Theoretical Background

Thus, Buddhism had a significant impact on the formation of business ethics in Japan. However, the influence of Buddhism on business development is not limited to only ethical requirements and justifications. Buddhist philosophy reflected on the principles, goals, and characteristics of management in the East (Dayong et al., 2020; Mikhaylov et al., 2018; Nyangarika et al., 2018; Danish et al., 2020, 2021; Lisin, 2020; An et al., 2021; Ivanyuk & Berzin, 2020; Ivanyuk & Levchenko, 2020; Ivanyuk et al., 2020; Ivanyuk, 2018; Ivanyuk & Soloviev, 2019; Conteh et al., 2021; Uyeh et al., 2021; Moiseev et al., 2021; Grilli et al., 2021).

It is becoming more and more clear speculative capital of global corporations creating responsible ethically oriented observational and regulatory governing body, ensuring openness and transparency the work of the world market, the mechanisms of formation of the structure global regulation. It is about using economically toolkit contained in religions as system ethical values. One example is the influence of Protestant ethics on a hundred the rise of capitalism (Denisova et al., 2019; Nyangarika et al., 2019a, b; Huang et al., 2021a, b; Mikhaylov, 2018a, b, 2022; Meynkhard, 2019, 2020; Mikhaylov et al., 2019).

The following example of the positive influence of religion on the energy management is linked to the Islamic banking system. Koran prohibits usury in all forms (Bhuiyan et al., 2021; Dong et al., 2021, Mikhaylov, 2021b; Liu et al., 2021, 2022; Saqib et al., 2021; Radosteva et al., 2018; Ranjbar et al., 2017; Rathnayaka et al., 2018; Sunchalin et al., 2019; Uandykova et al., 2020; Udalov, 2021; Yuksel et al., 2021a, b, c; Dorofeev, 2020; Mukhametov et al., 2021; Candila et al., 2021).

Buddhism and Confucianism had a great influence on the development of business ethics in Japan and other countries. In the twenty-first century, world capital moves to Southeast Asia. According to many researchers, the rapid economic breakthrough of the countries of Southeast Asia largely shares the ideology of Confucianism, which not only in China, but also in Korea, Vietnam, Singapore, Taiwan (Mikhaylov, 2018c; Mikhaylov et al., 2019; Melnichuk et al., 2020; Nie et al., 2020).

Confucianism is the basis of unity and integrity region, allowing it to withstand the expansion of real ideas. In the Confucian model of economics are not opinions and competition. The main values are maintaining order and respect for hierarchical relationships. At the same time, the weakness of Confucianism is the overwhelming dominance of ethics of duty and collectivism (An et al., 2019a, b; 2020a, b, c; Mikhaylov, 2019, 2021a; Mikhaylov & Tarakanov, 2020; Moiseev et al., 2020; Gura et al., 2020; Dooyum et al., 2020; Mikhaylov et al., 2020a, b, c, 2021a, b; Varyash et al., 2020; Zhao et al., 2021; An & Mikhaylov, 2020, 2021; Alwaelya et al., 2021; Yumashev & Mikhaylov, 2020; Yumashev et al., 2020; Mutalimov et al., 2021; Morkovkin et al., 2020a, b).

3 Significant Findings

The study uses data from Thomson Reuters about 164 countries of the first cycle ($n = 164$). The dataset includes GDP growth in Energy in 2018, quantity of Christians, Muslims, and Buddhists in each country. The regression analysis is a strong tool for the influence of one or more independent variables.

Comparison of regression results for the economic growth and level of religiousness is about 23%. But if clear any values the impact of religion on economic growth is higher (30–50%). In terms of a stable balance between collectivism, duty and freedom are of interest. It is the experience of Christianity and Buddhism. Underdeveloped economically part of this religious teaching is due to objective factors of its occurrence and specifically historical conditions. This gave rise to some scientists are critical of the potential in the field of economic development.

Comparison of regression results for the economic growth and Muslim population does not show the so stable link like before (about 10%). And this does not contribute to economic development, so as the goal of planning in a backward and poor country is it is customary to get rationalization from the people lifestyle and, in particular, encourage him to more stubbornly and purposeful work.

Buddhism, they say, is quite capable to directly and indirectly promote social and economic modernization and to perform constructive and progressive functions.

Studying these countries, researchers came to the conclusion that although Buddhist monks are not directly related to the leadership, the institution of monasticism through the transfer of value and norms still has a certain impact on unsocial-economic system, affecting both lifestyle and decision-making on production and consumption.

Therefore, a number of scientists note a positive role of Buddhism in shaping business ethics in Japan. The possibility of building a fundamentally new energy management and many specific recommendations of Buddhism can effectively but used in corporate activities. It is Buddhist economic philosophy that influenced Japanese economists, industrialists, entrepreneurs, and corporations effective leaders.

The advantage of socialism is that the ruler state or state controls or regulates distribution of the flow of wealth. Excessive taxation of capable employers and diligent workers in favor of the dysfunctional and the lazy are discouraged by individual initiatives, diligence and diligence. It limits human development potential.

Buddhist energy management guarantees a viable energy management development without damaging or exhausting the environment without harming human resources.

Malaysian Government Policy also borrowed some management techniques from Buddhist economic model and raised the energy management Malaysia's competitiveness. Many modern Buddhists believe that doing business does not contradict the commandments of Buddhism.

Having them, man can create much better than one who does not have these resources. Question with costs in the way we make money. To maintain a healthy attitude towards money, the main idea of a hundredth honesty of the way you earn them those. It is necessary to clearly understand the source of their income and do everything so that this source does not run out. The second principle is that we should enjoy give money, that is, learn to keep our thoughts and the body is in good health in the process of making money.

4 Conclusions and Discussion

Then aftersales months of spiritual quest in various Buddhist insistence, after which a completely different dexterity. The ethical and philosophical foundations of the development of modern economies also lie in the fact that a person must not waste to vain the forces bestowed by nature, and to live fully life in both the ordinary, mundane sense and immorally.

So, it becomes obvious that economic growth in itself is not able to lead to sustainable human community development. Many prominent scientists predict near-term depletion of natural resources, demographic crisis, and global environmental catastrophe in the event that the modern paradigm human development will continue over the next decades.

In this regard, scientists are increasingly turning their look at the old experience of the civilizations of the East, able to achieve harmony between man, society, and nature. The economic ideal of religions becomes more and more relevant.

The middle harmonious way between extremes of poverty and wealth, which suggests reasonable costs of energy and resources, on the one hand, and satisfactory human-flying results, on the other. And the valid but if you compare recommended by experts in the field global issues benchmarks optimal development modality for humanity in the twenty-first century with socially economic and sociocultural foundations of ancient civilizations flow, it turns out that many of these ideals have long been embodied in life.

Interest in Buddhist culture, philosophy, and ethics today is not only characteristic for eastern, but also for western people, some of whom become Buddhists. It should be noted that Japanese corporations are still practicing classes in Buddhist temples. Buddhist methods and approaches are universal and may be applicable to different situations and circumstances of modern life. Buddhist culture aimed at spiritual improvement of a person and building harmonious society, able to offer new approaches and models to all modern business world, where ethical standards and ideals are often violated.

As can be seen, religious beliefs are highly influential on energy investments. Energy is vital for a country (Zhong et al., 2020). Therefore, countries that meet their energy needs by importing them from other countries are at serious risk (Yüksel et al., 2020). In this context, countries urgently need to take actions to support their own energy resources (Li et al., 2020; Xie et al., 2021). In this framework, renewable energy investments can be increased. Thus, countries will be able to both have their own energy resources and produce the energy they need without harming the environment. In addition, countries may also attach importance to nuclear energy investments (Yuan et al., 2021; Meng et al., 2021). In this way, it will be possible for countries to produce uninterrupted energy (Yuksel et al., 2021; Yüksel & Çağlayan, 2020; Dinçer et al., 2020).

In summary, it is essential to increase energy investments for the sustainable economic development of countries. In this context, it is necessary to pay attention to all the factors that will increase the efficiency of these investments. Religious beliefs can also be very effective in this process. In this context, moral feelings play a very important role in the success of energy investments. For example, in order to increase the success of nuclear energy investments, necessary precautions should be taken against the risks in this process. Otherwise, the probability of an accident at nuclear power plants is quite high. This situation both increases the cost of investments and endangers people's lives. Therefore, it is very important to take the necessary precautions ethically during the installation process of nuclear power plants.

References

Alwaelya, S. A., Yousif, N. B. A., & Mikhaylov, A. (2021). Emotional development in preschoolers and socialization. *Early Child Development and Care, 190*(1717480).

An, J., & Mikhaylov, A. (2021). Analysis of energy projects financial efficiency and renewable energy generation in Russia. *Finance: Theory and Practice, 25*(5), 1–11.

An, J., & Mikhaylov, A. (2020). Russian energy projects in South Africa. *Journal of Energy in Southern Africa, 31*(3).

An, J., Mikhaylov, A., & Jung, S.-U. (2020a). The strategy of South Korea in the global oil market. *Energies, 13*(10), 2491.

An, J., Mikhaylov, A., & Kim, K. (2020b). Machine learning approach in heterogeneous group of algorithms for transport safety-critical system. *Applied Sciences, 10*(8), 2670.

An, J., Mikhaylov, A., & Moiseev, N. (2019a). Oil price predictors: Machine learning approach. *International Journal of Energy Economics and Policy, 9*(5), 1–6.

An, J., Mikhaylov, A., & Sokolinskaya, N. (2019b). Machine learning in economic planning: Ensembles of algorithms. *Journal of Physics: Conference Series, 1353*, 012126.
An, J., Mikhaylov, A., & Jung, S.-U. (2021). A linear programming approach for Robust network revenue management in the airline industry. *Journal of Air Transport Management, 91*(March 2021), 101979.
An, J., Mikhaylov, A., & Richter, U. H. (2020c). Trade war effects: Evidence from sectors of energy and resources in Africa. *Heliyon, 6*, e05693. https://doi.org/10.1016/j.heliyon.2020.e05693
Bhuiyan, M. A., An, J., Mikhaylov, A., Moiseev, N., & Danish, M. S. S. (2021). Renewable energy deployment and COVID-19 measures for sustainable development. *Sustainability, 13*, 4418. https://doi.org/10.3390/su13084418
Candila, V., Maximov, D., Mikhaylov, A., Moiseev, N., Senjyu, T., & Tryndina, N. (2021). On the relationship between oil and exchange rates of oil-exporting and oil-importing countries: from the great recession period to the Covid-19 era. *Energies, 14*, 8065.
Conteh, F., Takahashi, H., Hemeida, A. M., Krishnan, N., Mikhaylov, A., & Senjyu, T. (2021). Hybrid grid-connected renewable power generation for sustainable electricity supply in Sierra Leone: Case study Lungi Town. *Sustainability, 13*(18).
Danish, M. S. S., Bhattacharya, A., Stepanova, D., Mikhaylov, A., Grilli, M. L., Khosravy, M., & Senjyu, T. (2020). A systematic review of metal oxide applications for energy and environmental sustainability. *Metals, 10*(12), 1604.
Danish, M. S. S., Senjyu, T., Sabory, N. R., Khosravy, M., Grilli, M. L., Mikhaylov, A., & Majidi, H. (2021). A forefront framework for sustainable aquaponics modeling and design. *Sustainability, 13*, 9313. https://doi.org/10.3390/su13169313
Dayong, N., Mikhaylov, A., Bratanovsky, S., Shaikh, Z. A., & Stepanova, D. (2020). Mathematical modeling of the technological processes of catering products production. *Journal of Food Process Engineering, 43*(2), 1–9.
Denisova, V., Mikhaylov, A., & Lopatin, E. (2019). Blockchain infrastructure and growth of global power consumption. *International Journal of Energy Economics and Policy, 9*(4), 22–29.
Dinçer, H., Yüksel, S., Çağlayan, Ç., & Uluer, G. S. (2020). The contribution of nuclear energy investment on sustainable financial and economic development. *Journal of Financial Economics and Banking, 1*(1), 39–51.
Dong, B., Ikonnikova, I., Rogulin, R., Sakulyeva, T., & Mikhaylov, A. (2021). Environmental-economic approach to optimization of transport communication in megacities. *Journal of Environmental Science and Health, Part A, 56*(6), 660–666. https://doi.org/10.1080/10934529.2021.1913928
Dooyum, U. D., Mikhaylov, A., & Varyash, I. (2020). Energy security concept in Russia and South Korea. *International Journal of Energy Economics and Policy, 10*(4), 102–107.
Dorofeev, M. L. (2020). Impact of monetary policy on the level of economic inequality in the United States. *MGIMO Review of International Relations, 13*(5), 97–114. https://doi.org/10.24833/2071-8160-2020-5-74-97-114
Grilli, M. L., Valerini, D., Rizzo, A., Yilmaz, M., Song, C., Hu, G., Chierchia, R., Mikhaylov, A., & Rinaldi, A. (2021). A comparative study of the mechanical and tribological properties of thin Al2O3 coatings fabricated by atomic layer deposition and radiofrequency sputtering. *Physica Status Solidi (A), 218*(24), 1–25.
Gura, D., Mikhaylov, A., Glushkov, S., Zaikov, M., & Shaikh, Z. A. (2020). Model for estimating power dissipation along the interconnect length in single on-chip topology. *Evolutionary Intelligence*, s12065.
Huang, Y., Masrur, H., Shigenobu, R., Hemeida, A. M., Mikhaylov, A., & Senjyu, T. (2021a). A comparative design of a campus microgrid considering a multi-scenario and multi-objective approach. *Energies, 14*, 2853. https://doi.org/10.3390/en14112853
Huang, Y., Yona, A., Takahashi, H., Hemeida, A. M., Mandal, P., Mikhaylov, A., Senjyu, T., & Lotfy, M. E. (2021b). Energy management system optimization of drug store operate electric vehicles charging station. *Sustainability, 14*, 1217836.

Ivanyuk, V., & Berzin, D. (2020). An algorithm for constructing an efficient investment portfolio. In R. Silhavy, P. Silhavy, & Z. Prokopova (Eds.), *Software engineering perspectives in intelligent systems. CoMeSySo 2020. Advances in intelligent systems and computing* (Vol. 1294). Springer. https://doi.org/10.1007/978-3-030-63322-6_39

Ivanyuk, V., & Levchenko, K. (2020). Intelligent methods for predicting financial time series. In R. Silhavy, P. Silhavy, & Z. Prokopova (Eds.), *Software engineering perspectives in intelligent systems. CoMeSySo 2020. Advances in intelligent systems and computing* (Vol. 1294). Springer. https://doi.org/10.1007/978-3-030-63322-6_41

Ivanyuk, V., Sunchalin, A., & Sunchalina, A. (2020). Development of an intelligent ensemble forecasting system. In R. Silhavy, P. Silhavy, & Z. Prokopova (Eds.), *Software engineering perspectives in intelligent systems. CoMeSySo 2020. Advances in intelligent systems and computing* (Vol. 1294). Springer. https://doi.org/10.1007/978-3-030-63322-6_40

Ivanyuk, V. (2018). *Econometric forecasting models based on forecast combination methods*. Proceedings of 2018 11th International Conference Management of Large-Scale System Development, MLSD 2018. https://doi.org/10.1109/MLSD.2018.8551825.

Ivanyuk, V., & Soloviev, V. (2019). *Efficiency of neural networks in forecasting problems*. Proceedings of 2019 12th International Conference Management of Large-Scale System Development; MLSD 2019. https://doi.org/10.1109/MLSD.2019.8911046.

Li, X., Zhu, S., Yüksel, S., Dinçer, H., & Ubay, G. G. (2020). Kano-based mapping of innovation strategies for renewable energy alternatives using hybrid interval type-2 fuzzy decision-making approach. *Energy, 211*, 118679.

Lisin, A. (2020). Valuation of the activities of foreign banks in the Russian banking sector. *Orbis, 15*(45), 53–63. http://www.revistaorbis.org.ve/pdf/45/art5.pdf

Liu, Z.-J., Panfilova, E., Mikhaylov, A., & Kurilova, A. (2021). COVID-19 crisis impact on the stability between parties in crowdfunding and crowdsourcing. *Wireless Personal Communications, 8932*. https://doi.org/10.1007/s11277-021-08932-z

Liu, Z.-J., Panfilova, E., Mikhaylov, A., & Kurilova, A. (2022). Assessing stability in relationship between parties in crowdfunding and crowdsourcing projects during the COVID-19 crisis. *Journal of Global Information Management, 30*(2), 1.

Melnichuk, A. V., Melnichuk, Y. A., Bocharov, A. V., Stepanova, D. I., & Kurbanov, S. A. (2020). Constitutional foundations of the functioning of local self-government. *Revista Inclusiones., 7*, 337–346.

Meng, Y., Dinçer, H., & Yüksel, S. (2021). Understanding the innovative developments with two-stage technology S-curve of nuclear energy projects. *Progress in Nuclear Energy, 140*, 103924.

Meynkhard, A. (2020). Effect of Bitcoin volatility on altcoins pricing. *Advances in Intelligent Systems and Computing, 1294*, 652–664. https://doi.org/10.1007/978-3-030-63322-6_55

Meynkhard, A. (2019). Fair market value of bitcoin: halving effect. *Investment Management and Financial Innovations, 16*(4), 72–85. https://doi.org/10.21511/imfi.16(4).2019.07

Mikhaylov, A., Danish, M. S. S., & Senjyu, T. (2021a). New stage in evolution of cryptocurrency market: Analysis by hurst method. In H. Dincer & S. Yuksel (Eds.), *Strategic outlook in business and finance innovation: Multidimensional policies for emerging economies* (pp. 35–45). Emerald Publishing Limited. https://doi.org/10.1108/978-1-80043-444-820211004

Mikhaylov, A. (2018a). Volatility spillover effect between stock and exchange rate in oil exporting countries. *International Journal of Energy Economics and Policy, 8*(3), 321–326.

Mikhaylov, A. (2018b). Pricing in oil market and using probit model for analysis of stock market effects. *International Journal of Energy Economics and Policy, 8*(2), 69–73.

Mikhaylov, A. (2018c). The yield of the carry trade strategy. *Finance: Theory and Practice, 22*(3), 52–63. https://doi.org/10.26794/2587-5671-2018-22-3-52-63

Mikhaylov, A. (2019). Oil and gas budget revenues in Russia after crisis in 2015. *International Journal of Energy Economics and Policy, 9*(2), 375–380.

Mikhaylov, A. (2020a). Cryptocurrency market analysis from the open innovation perspective. *Journal of Open Innovation: Technology, Market, and Complexity, 6*(4), 197. https://doi.org/10.3390/joitmc6040197

Mikhaylov, A. (2020b). Geothermal energy development in Iceland. *International Journal of Energy Economics and Policy, 10*(4), 31–35.

Mikhaylov, A. (2020c). Cryptocurrency market development: hurst method. *Finance: Theory and Practice, 24*(3), 81–91.

Mikhaylov, A. (2021a). Development of Friedrich von Hayek's theory of private money and economic implications for digital currencies. *Terra Economicus, 19*(1).

Mikhaylov, A. (2021b). Lichens as indicators of atmospheric pollution in urban ecosystems. *Israel Journal of Ecology & Evolution, 67*(1–2), 60–68. https://doi.org/10.1163/22244662-bja10016

Mikhaylov, A. (2022). Efficiency of renewable energy generation in Russia. *Annals of the Brazilian Academy of Sciences, 94*(3).

Mikhaylov, A., & Tarakanov, S. (2020). Development of Levenberg-Marquardt theoretical approach for electric network. *Journal of Physics: Conference Series, 1515*, 052006.

Mikhaylov, A., Burakov, D., & Didenko, V. (2019). Relationship between oil price and macroeconomic indicators in Russia. *Finance: Theory and Practice, 23*(2), 105–116. https://doi.org/10.26794/2587-5671-2019-23-2-105-116

Mikhaylov, A., Sokolinskaya, N., & Nyangarika, A. (2018). Optimal carry trade strategy based on currencies of energy and developed economies. *Journal of Reviews on Global Economics, 7*, 582–592.

Mikhaylov, A., Yumashev, A., & Kolpak, E. (2021b). Quality of life, anxiety and depressive disorders in patients with extrasystolic arrhythmia. *Archives of Medical Science, 17*(1). https://doi.org/10.5114/aoms.2020.101359

Moiseev, N., Mikhaylov, A., Varyash, I., & Saqib, A. (2020). Investigating the relation of GDP per capita and corruption index. *Entrepreneurship and Sustainability Issues, 8*(1), 780–794.

Moiseev, N., Sorokin, A., Zvezdina, N., Mikhaylov, A., Khomyakova, L., & Danish, M. S. S. (2021). Credit risk theoretical model on the base of DCC-GARCH in time-varying parameters framework. *Mathematics, 9*, 2423. https://doi.org/10.3390/math9192423

Morkovkin, D., Lopatkin, D., Sadriddinov, M., Shushunova, T., Gibadullin, A., & Golikova, O. (2020a). Assessment of innovation activity in the countries of the world. *E3S Web of Conferences, 157*, 04015. https://doi.org/10.1051/e3sconf/202015704015

Morkovkin, D. E., Gibadullin, A. A., Kolosova, E. V., Semkina, N. S., & Fasehzoda, I. S. (2020b). Modern transformation of the production base in the conditions of Industry 4.0: problems and prospects. *Journal of Physics*: Conference Series, *1515*, 032014.

Mukhametov, A., Bekhorashvili, A., Avdeenko, A., & Mikhaylov, A. (2021). The impact of growing legume plants under conditions of Biologization and Soil Cultivation on Chernozem fertility and productivity of rotation crops. *Legume Research, 44*, 12.

Mutalimov, V., Kovaleva, I., Mikhaylov, A., & Stepanova, D. (2021). Assessing regional growth of small business in Russia. *Entrepreneurial Business and Economics Review, 9*(3), 119–133. https://doi.org/10.15678/EBER.2021.090308

Nie, D., Panfilova, E., Samusenkov, V., & Mikhaylov, A. (2020). E-Learning financing models in Russia for sustainable development. *Sustainability, 12*(11), 4412.

Nyangarika, A., Mikhaylov, A., & Richter, U. (2019a). Influence oil price towards economic indicators in Russia. *International Journal of Energy Economics and Policy, 9*(1), 123–129.

Nyangarika, A., Mikhaylov, A., & Richter, U. (2019b). Oil price factors: Forecasting on the base of modified auto-regressive integrated moving average model. *International Journal of Energy Economics and Policy, 9*(1), 149–159.

Nyangarika, A., Mikhaylov, A., & Tang, B.-J. (2018). Correlation of oil prices and gross domestic product in oil producing countries. *International Journal of Energy Economics and Policy, 8*(5), 42–48.

Radosteva, M., Soloviev, V., Ivanyuk, V., & Tsvirkun, A. (2018). Use of neural network models in market risk management. *Advances in Systems Science and Applications, 18*(2), 53–58. https://doi.org/10.25728/assa.2018.18.2.582

Ranjbar, O., Chang, T., Nel, E., & Gupta, R. (2017). Energy consumption and economic growth nexus in South Africa: Asymmetric frequency domain approach. *Energy Sources, Part B: Economics, Planning, and Policy, 12*(1), 24–31.

Rathnayaka, R. K. T., Seneviratna, D. M. K., & Long, W. (2018). The dynamic relationship between energy consumption and economic growth in China. *Energy Sources, Part B: Economics, Planning, and Policy, 13*(5), 264–268.

Saqib, A., Chan, T.-H., Mikhaylov, A., & Lean, H. H. (2021). Are the responses of sectoral energy imports asymmetric to exchange rate volatilities in Pakistan? Evidence from recent foreign exchange regime. *Frontiers in Energy Research, 9*, 614463. https://doi.org/10.3389/fenrg.2021.614463

Sunchalin, A. M., Kochkarov, R. A., Levchenko, K. G., Kochkarov, A. A., & Ivanyuk, V. A. (2019). Methods of risk management in portfolio theory. *Espacios, 40*(16), 8.

Uandykova, M., Lisin, A., Stepanova, D., Baitenova, L., Mutaliyeva, L., Yuksel, S., & Dincer, H. (2020). The social and legislative principles of counteracting ransomware crime. *Entrepreneurship and Sustainability Issues, 8*(2), 777–798.

Udalov, I. (2021). The transition to renewable energy sources as a threat to resource economies. *International Journal of Energy Economics and Policy, 11*(3), 1–6.

Uyeh, D. D., Asem-Hiablie, S., Park, T., Kim, K.-M., Mikhaylov, A., Woo, S., & Ha, Y. (2021). Could Japonica rice be an Alternative variety for increased global food security and climate change mitigation? *Food, 10*, 1869. https://doi.org/10.3390/foods10081869

Varyash, I., Mikhaylov, A., Moiseev, N., & Aleshin, K. (2020). Triple bottom line and corporate social responsibility performance indicators for Russian companies. *Entrepreneurship and Sustainability Issues, 8*(1), 313–331.

Xie, Y., Zhou, Y., Peng, Y., Dinçer, H., Yüksel, S., & an Xiang, P. (2021). An extended Pythagorean fuzzy approach to group decision-making with incomplete preferences for analyzing balanced scorecard-based renewable energy investments. *IEEE Access, 9*, 43020–43035.

Yuan, G., Xie, F., Dinçer, H., & Yüksel, S. (2021). The theory of inventive problem solving (TRIZ)-based strategic mapping of green nuclear energy investments with spherical fuzzy group decision-making approach. *International Journal of Energy Research.*

Yuksel, S., Khomyakova, L., & Mikhaylov, A. (2021). Energy center selection in G7 Industry with Fuzzy MOORA. In S. Yüksel & H. Dinçer (Eds.), *Handbook of research on strategic management for current energy investments (pp. 87–106)*. IGI Global. https://doi.org/10.4018/978-1-7998-8335-7.ch006

Yüksel, S., Mikhaylov, A., & Ubay, G. G. (2021a). Factors Causing Delay in the Installation of Nuclear Power Plants. In S. Yüksel & H. Dinçer (Eds.), *Strategic approaches to energy management. Contributions to management science*. Springer. https://doi.org/10.1007/978-3-030-76783-9_7

Yüksel, S., & Çağlayan, Ç. (2020). The factors affecting the investors' decisions: A study on nuclear energy investments. *Economics Literature, 2*(2), 177–185.

Yüksel, S., Dinçer, H., & Uluer, G. S. (2020). The role of technological development on renewable energy usage: An econometric analysis for G7 countries. In *Handbook of research on sustainable supply chain management for the global economy* (pp. 136–153). IGI Global.

Yüksel, S., Dinçer, H., Çağlayan, Ç., & Uluer, G. S. (2021b). Determining the optimal financial strategies for nuclear energy companies. In *Financial strategies in competitive markets* (pp. 1–16). Springer.

Yüksel, S., Mikhaylov, A., Ubay, G. G., & Uyeh, D. D. (2021c). The role of hydrogen in the black sea for the future energy supply security of Turkey. In S. Yüksel & H. Dinçer (Eds.), *Handbook of research on strategic management for current energy investments (pp. 1–15)*. IGI Global. https://doi.org/10.4018/978-1-7998-8335-7.ch001

Yumashev, A., & Mikhaylov, A. (2020). Development of polymer film coatings with high adhesion to steel alloys and high wear resistance. *Polymer Composites, 41*(7), 2875–2880.

Yumashev, A., Ślusarczyk, B., Kondrashev, S., & Mikhaylov, A. (2020). Global indicators of sustainable development: Evaluation of the influence of the human development index on consumption and quality of energy. *Energies, 13*, 2768.

Zhao, Q., Cherkasov, S., Avdeenko, A., Kondratenko, L., & Mikhaylov, A. (2021). Integral estimate of the added value of agricultural crops in the synergetic agriculture on the example of Vicia faba (Vicia faba L.). *Polish Journal of Environmental Studies, 30*(5), 1–9. https://doi.org/10.15244/pjoes/131949

Zhong, J., Hu, X., Yüksel, S., Dinçer, H., & Ubay, G. G. (2020). Analyzing the investments strategies for renewable energies based on multi-criteria decision model. *IEEE Access, 8*, 118818–118840.

Implications of Energy Subsidies from Economic Standpoint

Cansın Kemal Can

1 Introduction

The substantial industrial progress and the massive scale of production achieved in the last two centuries could not have been possible without utilizing the enormous energy sources of our planet. Nevertheless, notwithstanding its outstanding contribution to economic welfare, fossil energy sources have brought about severe challenges on social and economic grounds. Thus, in addition to promoting economic growth in every country, contemporary international society confronts an additional task of minimizing the side effects of intense energy consumption.

To maximize the economic outcome of energy consumption, the governments impose subsidy policies in several forms to promote production in certain industries and to make the energy sources available at lower prices for the consumers. Yet, despite their positive impact on the economic performance, the energy subsidies also impair the economic and social posture of the country in numerous ways including income inequality, forgone revenues, depressed growth, the deteriorating balance of payments, profligacy in energy consumption. As a result of these detrimental effects of subsidies, energy consumption culminates in a deterioration in the economy rather than the intended targets regarding economic development. Despite their theoretical growth-inducing effects, the energy subsidies mostly give rise to economic disruption due to their unintended side effects. This controversy entails a firm understanding of those unpleasant repercussions of subsidies to shun the adverse outcomes through rigorous reform strategies. This study aims to elaborate on the consequences and the implications of energy subsidies to offer caveats for reducing the counterproductive impact thereof. To accomplish this objective, in the following section

C. K. Can (✉)
Istanbul Medeniyet University, Istanbul, Turkey
e-mail: kemal.can@medeniyet.edu.tr

© The Author(s), under exclusive license to Springer Nature Switzerland AG 2022
H. Dinçer, S. Yüksel (eds.), *Sustainability in Energy Business and Finance*, Contributions to Finance and Accounting,
https://doi.org/10.1007/978-3-030-94051-5_18

some theoretical considerations are presented followed by primary and secondary economic implications of energy subsidies on several aspects of the economy such as output, public finance, balance of payments, income inequality, etc. Then, some strategies to mitigate the adverse side effects of energy subsidies are discussed before concluding with final remarks regarding current status of energy subsidy reform.

2 Theoretical Considerations Regarding the Implications of Energy Subsidies

Energy subsidies can succinctly be defined as the government regulations on the energy sources striving to cut the production costs of producers and the energy expenditures of the households. Also, in some cases, the energy subsidies might be designed to promote the income of energy suppliers. Mostly, consumer subsidies occur when the price is set below a threshold level, which is generally equal to the supply cost. The supply cost of energy oftentimes takes transportation and distribution costs into account as well. The producer subsidies, on the other hand, involve aiding the supplier firms through tax reductions or direct price setting.

In general, from consumption viewpoint, energy subsidies are pursued to protect the accession possibilities of the poorer segments of society to energy resources. In addition, it is also designed to encourage the use of cleaner and more effective energy sources and thereby increase the living standards of the public as a whole. By subsidizing the use of natural gas, for instance, the government aims to eliminate other energy sources such as firewood or kerosene which are comparatively more primitive, dirty and cumbersome. Furthermore, these energy sources lead to deforestation and air pollution which are additional externalities for the society.

According to Robinson and Mitchell (2020), 69% of the total energy subsidies are devoted to oil which is followed by gas subsidies which is only 17%. The distribution of energy subsidies with respect to resource types implies that that the oil is the most heavily subsidized energy source since it is the most critical and productive natural resource with its widespread use in each segment of the economy. Another reason for the existence of intensive oil subsidies is the unending fluctuations in the international oil prices. Every country strives to preserve their domestic economy from the volatilities in the oil prices which most of the time occur abruptly. From a macroeconomic viewpoint, by implementing the energy subsidies, the government purposes to cushion the economy from fluctuations in the international market. Energy prices are among the most important factors for the stability of the economy; therefore, the governments use energy subsidies to protect both the consumers and the producers from the unpleasant adverse effects of the swings in energy prices. As such, the government preserves the purchasing power of the consumers while keeping the competitiveness of the producers at a certain level. Besides, in many societies, energy subsidies serve the function of the social safety net by facilitating income distribution while protecting the poor (Edgar Cooke et al., 2016). Thus, they play a key role in the establishment of well-settled wealth distribution and social justice in the country.

Nevertheless, notwithstanding their positive aspects, in practice, oftentimes the energy subsidies fail to reach their intended positive goals and generate a fiscal burden on the country's budget generating a substantial amount of deadweight loss. In fact, the loss in efficiency is the primary reason for the loss in social welfare. In particular, producer subsidies generally lead to overproduction while consumer subsidies bring about wasted energy through profligacy in consumption (Parry et al., 2014). Thus, energy subsidies mostly cause a mismatch between the demand for and the supply of energy sources in the country. Just as in the case of other types of government interventions, the end result of the energy subsidies is chiefly distortionary for the economy and society.

3 Economic Implications of Energy Subsidies

Now that we have explored theoretical considerations for energy subsidies, we can consider the economic implications of energy subsidies which are technically reverse taxation. As we mentioned earlier, despite their theoretical economic benefits, the energy subsidies mostly impair the economic activities. Basically, energy subsidies beget consequences on the economy through three channels: Output level, public finances and balance of payments.

3.1 Implications on Output

There are several ways through which energy subsidies affect the size of the output in the economy. Firstly, the output level is impacted by the energy subsidies because the energy subsidies lead to a reduction in energy investments. Since the competition is reduced by subsidies towards state-owned enterprises, the new firms have smaller incentives in investing in those markets especially in developing economies. Lack of competition in those sectors lead to efficiency losses which cause higher-than-normal prices in the energy markets of those countries (Barkhordar et al., 2018). Also, the lack of private investment in those countries brings about reduced energy capacity, which eventually causes energy shortages. In modern economies, such shortages are disruptive for growth since production is mostly interrupted due to the halts in energy production (Bazilian & Onyeji, 2012).

Secondly, the market efficiency is distorted by the aforementioned reduction in the level of competitiveness in the market. By altering the levels of supply and demand via market manipulations, the subsidies effectively determine the level of production and consumption along with the equilibrium price in the market, which is a genuine violation of the market laws of supply and demand. Under subsidy schemes, the price is always lower than the market-clearing price and the quantity is always higher than the equilibrium level. The point here is that the disequilibrium established by the energy subsidies permeates all sectors of the economy very swiftly leading to an eventual inefficient allocation in the entire economy (Wattanakuljarus, 2021).

Also, the subsidized energy source will be used abusively compared to alternative resources which is another source of disruption for competitiveness. Once the resource allocation is impaired through energy subsidies, the entire economic activities plunge due to the mismatch in the supply and demand in each sub-sector of the economy. Thus, it is clear that energy subsidies lead to malfunctioning market mechanisms and the consequent economic downturn is inevitable when the energy subsidies are used excessively.

The final way the energy subsidies affect the economic activities is the crowding-out effect of the public financial resources spent or forgone for the implementation of the subsidies. The governments finance energy subsidies either directly through budgetary transfers or indirectly through relinquished tax revenues. In other words, there are explicit and implicit ways of financing energy subsidies from the viewpoint of public finances. The implicit financing leads to a suboptimal level of taxation since the government forgoes a portion of accruing tax revenue. No matter which types of financing are used by the government to finance the energy subsidies, the fiscal space shrinks inefficiently because the financial resources which could have been used in more efficient and effective projects are diverted to financing energy subsidies which is by nature a cause of market failure.

3.2 Implications on Public Finances

Public finances are the second channel through which energy subsidies affect the economy. The sole financier of the energy subsidies is oftentimes the government. Thus, any surging trend in the cost of energy adversely impacts the government budget. Any rise in the international oil prices, for instance, brings about a substantial fiscal burden on the economies. In the case of developing countries, the fiscal space is mostly insufficient to service the fiscal costs of energy subsidies; therefore, mounting energy subsidy costs consequently lead to a surge in public debt (Acharya & Sadath Anver, 2017). In this case, the fiscal cost of energy subsidies gets even higher due to compounding interest on the accumulated debt which was incurred to finance the energy subsidies. This prevents the government from making more beneficial and efficient spending since the fiscal space is reduced due to interest expenditures. Also, some energy subsidies are implemented in the form of tax reductions or exemptions. If this type of subsidy is preferred, then government forgoes a portion of its expected revenue in the upcoming periods. This generates another source of fiscal burden on the public finances since the falling tax revenues have to be compensated by other sources of financing (Vagliasindi, 2013).

As in the case of increasing energy costs, forgone tax revenues also lead to excessive borrowing, which is detrimental for developing countries since they are mostly unable to borrow under good conditions such as long maturities and low-interest rates. Those countries, mostly suffer from high levels of risk premiums due to heightened risk of default. The low credibility rating of the sovereign creates another challenge for the compensation of the financing needs occurring as a result of energy subsidies in the form of reduced tax revenues. Moreover, due mostly to politico-economic reasons, the governments attempt the preserve the level of fuel prices stable even if the international oil prices move upwards (Breisinger et al., 2019). In order to trim the excessive upward movements in the fuel prices, the governments oftentimes have to reduce the tax burden on the fuel prices to maintain a stable price level. Nevertheless, these price stabilization policies lead to a misallocation of public funds since a substantial amount of budgetary funds which could have been diverted to more productive areas are used for this purpose. Hence, the surge in international oil prices brings about a decline in the efficiency of the overall economy especially in developing countries through budgetary spending (Coxhead & Grainger, 2018).

According to Granado et al. (2010), the energy subsidies lead to allocative problems since subsidized products are overused while the others are utilized below optimal levels. The uncompetitive and suboptimal resource allocation brings about large deadweight losses in the economy which in turn rises the fiscal bill of the energy subsidies. The reason is that the uncompetitive and inefficient markets do not perform at their profit maximizing levels and thereby do not fully utilize their productive capacity. Therefore, the government tax revenues fall dramatically due to the poor performance of the economy.

3.3 *Implications on Balance of Payments*

Energy subsidies also pose a challenge for the economic activities through balance of payments. The strength of this type of effect is even severer for oil-importing countries since, under energy subsidy reform, the quantity of energy used in the country is greater than the regime with no subsidy. In this case, the public funds are transferred to oil-producing countries and the domestic fiscal space shrinks, which potentially leads to less government spending on social welfare. This scenario exacerbates the public balances as well as macroeconomic balances through heightened imports and increasing reimbursements. In the oil-exporting countries, on the other hand, the energy subsidies lead to overconsumption of energy sources. Thus, these countries cannot fully benefit from the alleviating effects of increasing oil prices on their balance of payments. In order to meet the surging domestic demand for energy (thanks to energy subsidies), they need to reduce their energy exports even in the case of mounting oil prices in the international markets. This reduction brings about a loss in efficiency in the overall economy along with an improper allocation of resources as well as deterioration in the balance of payments. The financial loss from the decline in the energy export results in higher taxes and/or borrowing, which in turn worsens the public financial balances even further.

4 The Collateral Economic Implications of Energy Subsidies

Along with their direct effects on the overall economy, energy subsidies can also have indirect effects on the economic performance. The collateral effects of the energy subsidies occur due largely to the negative externalities caused by the implementation of those subsidies (Coady et al., 2015). As mentioned in the earlier sections, energy subsidies lead to lower prices, which in turn increases the quantity of energy consumed. The excessive consumption of the subsidized energy gives rise to overcombustion which is a substantial source of CO_2 emissions and thereby leads to drastic increases in the level of air pollution. In order to reduce the CO_2 emissions, the governments have to incur several other costs in addition to the heightened health and environmental expenditures. Thus, while subsidizing the energy sources, the governments effectively increase their total cost beyond the nominal cost of subsidies.

Also, the rate of traffic congestion in heavily subsidized countries is remarkably high, which brings about an inefficient resource allocation and extravagancy in terms of time and energy. Moreover, the fuel price reduction through energy subsidies leads to an increase in the number of cars on the road which in turn give rise to more fatalities and accidents which are additional sources of fiscal burden on the budget. In addition, if one type of source of energy is subsidized heavily, then it is preferred in remarkably high amounts compared to other and potentially cleaner energy sources due to profit maximization motivations. This situation creates a disincentive for the private sector for investing in clean energy resources for the same reasons. Thus, the new investment is diverted to energy-intensive technologies which are more costly to the economy and the environment. As result, the efficient and cleaner innovations do not come up, which impels the government to subsidize the old and costly technologies even further. In other words, energy subsidies generate their own vicious circle of ineffectiveness.

Along with these aspects, at the micro level, the energy subsidies also affect the economies through inequalities in income among the social groups in the society. The reason is that the benefits of the energy subsidies are not equally shared by the income groups. The low-income group mostly benefits from the energy subsidies indirectly while the higher income groups utilize the same energy subsidies mostly directly and become the sole benefiter of the subsidies. For instance, the price reduction in fuel might lead to a decline in the prices of other products such as public transportation, agricultural products, etc. Thus, low-income groups mostly reap benefits through these secondary price reduction channels since they do not consume the fuel directly unlike high-income society. In other words, the upper-income group section of the population generally consumes more energy compared to low-income groups which enable them to reap the full benefit of the energy subsidies (Granado et al., 2010). In Sub-Saharan Africa, for instance, a large portion of the population does not even have a proper link to the electricity grid. Thus, these people can only benefit from the energy subsidies in a collateral manner (Alleyne, 2013).

In sum, the energy subsidies are mostly not well-designed in terms of sharing the benefits. Since the subsidies are designed on a quantity-based approach, the more a household or a firm consumes the subsidized energy, the more benefit they reap from the economic facilities of the subsidies. The ultimate result of this process is a biased income distribution which is damaging for the long-run growth path of the economy. The income gap which occurs as a result of energy subsidies entails a larger amount of social transfers in each round which is an additional source of fiscal burden on the budget. In other words, once the energy subsidies are launched, the procedure automatically creates a compounding cost circle for the public budget (Schaffitzel et al., 2020).

5 Strategies to Mitigate the Adverse Economic Implications of Energy Subsidies

In the previous section, we have discussed how the energy subsidies lead to sizeable deteriorations in the fiscal balances, the overall economic performance along with societal and environmental damages. At this point, the obvious question is as to whether it is possible to alleviate the negative externalities and damages caused by the energy subsidies. The excessive fiscal costs associated with ongoing energy subsidies can only be reduced by implementing a certain degree of amendments and reforms in the scale and the scope of the energy policies. Nevertheless, these types of reforms in the energy subsidies entail detailed analysis of the primary and the secondary costs of energy subsidies and only then the energy subsidies can be anchored commensurate with the incurred costs by each segment of the society (Clements et al., 2013).

In practice, however, several hindrances mostly obscure the proper implementation of these types of reforms regarding the energy subsidies. One of the most prominent obstacles to a proper energy reform is the information asymmetries among the government and society. In particular, in most cases, the households are not well informed about the total cost of energy subsidies incurred by the government's budget (Rentschler & Bazilian, 2017). This information is usually not shared with society and therefore people are unaware of the extravagantly diverted funds through energy subsidies. Consequently, society is unable to ponder on the potential alternative uses of those funds such as education, health, infrastructure which can directly contribute to their welfare. Therefore, the asymmetric information brings about a challenge for a proper design of the energy subsidies in many countries. From economical perspective, these politico-economic trend leads to lack of transparency in public account and consequently brings about increasing fiscal costs and declining public welfare (Inchauste & Victor, 2017).

In addition to the information asymmetries, the second hindrance to the mitigation of adverse effects of energy subsidies is the lack of sovereign credibility, especially in developing countries. The countries in which lobbying and logrolling are prevalent are largely shaped by political mistrust since the society does not have confidence in the way the government will be using the saved funds from the restructured energy subsidies (Kyle, 2018). According to Alesina et al. (2008), especially in developing countries, society demands more transfer payments when the public funds rise since they believe that the government will use the raised funds in its own interest in a corrupt way rather than promoting public welfare. Hence, since society believes that some interest groups will reap the full benefits of energy subsidy reform, it is challenging to implement a full-fledged subsidy reform in those countries.

Moreover, while designing the energy subsidy plans, it is also essential to foresee the potential impact of the subsidies on income distributions so as to avoid distortions in income inequalities. As we mentioned earlier, the energy subsidies are mostly utilized by the high-income group of the society since they consume larger amounts of the subsidized product. This situation brings about a welfare loss for the low-income group of the society. Therefore, the energy subsidies in a way generate an income transfer from low to upper-income group which renders the poor segment of the society more vulnerable to the price fluctuations (Andriamihaja & Vecchi, 2007). Nevertheless, in practice, it is quite a challenge for the authorities to adjust the design of the energy subsidies to incorporate rationing among different income groups of the society. Also, the subsectors of the society which potentially face larger losses due to energy subsidy reforms are mostly well-organized and are experienced in cooperative acting such as unions, commercial interest groups, etc. Thus, in the case of an energy subsidy reform, they can withstand the government in a collective and powerful manner unlike households with loose formation. Therefore, even though the government needs to find a midway between the interests of households and the other interest groups while designing energy subsidies, the energy subsidies, per se, are burdensome for the government budget and tailoring these subsidies for reducing the income inequality generating effects makes them even costlier for the fiscal balances. Hence, the governments are mostly impelled by the interest groups to overlook the income gap generating effects of energy subsidies.

Besides, from a macro-fiscal perspective, redesigning an energy subsidy policy causes a rise in expectations regarding inflation and price fluctuations since removing subsidies increases the energy prices and since energy is the most important input for many industries, the inflationary expectations soar in the entire economy swiftly. As in the case of welfare-preserving policies for the poor, the government needs to apply macroeconomic policies to oversee the price fluctuations which occur as an outcome of the energy subsidy reform (Sovacool, 2017). The potential short- and medium-term effects of each policy alternative should be assessed in detail. The short-term effects on inflation mostly depend on the share of the specified product in the price index basket while the medium effects are influenced by the demand pressures and expectations. The optimal policy choice entails a rigorous evaluation of the ultimate temporal costs associated with the energy subsidies. Otherwise, the inflation will increase in a propagative manner depending on the size of the response by the CB.

There are several factors for determining the extent to which government subsidies give rise to inflation in the economy. The expectations of the market players, the coverage of the energy reform, the prevalence of the subsidized product in the market as input are the most prominent ones among many others. If the energy prices are represented by a large share in the price index basket and if the energy is used intensively in the country, then the inflationary effect of energy subsidy will be remarkably higher compared to other countries. In addition, the credibility rating of the central bank plays a key role in the strength of inflationary forces of energy subsidies. If the CB exhibited poor performance in controlling the inflation in the past, the households and the firms will adjust their expectations accordingly and alter their positions to avoid the unpleasant effects of looming inflationary environment which will further contribute to the realizations of higher inflation in the country. In the case of low CB credibility, even a one-shot price hike might result in a severe upward movement in inflation. Thus, a controlled a reasonable reaction by the CB is more effective than amplified excessive reactions which aim to stabilize the price fluctuations in a very limited time interval. The ultimate target of the CB policies should be to reduce the permeative effect of an abrupt price increase in the subsidized sector and protect stability.

The link between subsidy removal and price volatility also leads to heightened concerns about the competitiveness of certain sectors of the economy in the international market. In other words, if the energy subsidy is removed or redesigned in one country, the companies in the same sector from the countries with existing subsidies become comparatively competitive which reduces the profitability of the companies where the energy subsidies are removed. Thus, the authorities need to consider the trade-off between commercial loss of the private sector and income inequality enlargement due to the implementation of the energy subsidies (Töpfer, 2004).

Thus, in order to shun those unpleasant scenarios, the energy subsidies should be designed with a long-term goal in mind. During the planning stage of the subsidies, the long-term objectives need to be set in a clear-cut manner to reduce the trade-off assessments in the following stages of the energy subsidies. Besides, the alternative scenarios regarding the potential costs incurred by society and the economy need to be forecasted before the energy subsidies are designed. According to the forecast results, to avoid potential conflicts, the subsidies should be redesigned to minimize the impact of the incipient social crisis in the country. Also, after rigorously forecasting the costs associated with the ongoing subsidies, the subsidy plan needs to be approved by the parties involved in the subsidy such as households, enterprises, unions, etc. In this manner, the above-mentioned conflicts between interest groups can be overcome with negotiations before the energy subsidies are put in practice. Lack of proper communication among government and interest groups might bring about strong withstand in the later stages of the process. Hence, the information about the benefits and the costs of the energy subsidies needs to be publicly available in order for the groups to elaborate on the opportunity costs of subsidy reforms. Thus, it is clear that transparency is a key element in the performance of energy subsidies.

In addition, to achieve the long-term targets of the energy subsidies, it is more appropriate to implement back-loading price adjustments rather than front-loading strategies. The former type of pricing involves a gradual-alteration in the price of the subsidized product, whereas the latter refers to a case where the prices are changed abruptly. In the case of front loading, it is possible to receive resistance from interest groups since no time is allowed for adjustment. In the back-loading case, however, the involved parties are able to modify their position and optimize their consumption behaviour accordingly which is less distortionary for their budget. It is worthwhile to stress that it is crucial for the government to implement measures to preserve the welfare of the low-income group against the adverse effects of energy subsidies. Despite the fact that back-loading is an effective policy in this regard, it is by no means sufficient and requires additional measures to be taken. The additional measures to compensate the welfare loss of the poor might include, vouchers, transfers, specific tax cuts, etc., depending on the country-specific conditions.

6 Final Remarks

The energy subsidies incentivize the overuse of energy sources extravagantly. The subsidies targeting the producers lead to inefficiencies in the overall economy while the consumer subsidies give rise to income inequalities since the subsidies are mostly reaped by the wealthier segment of the society. Thus, it is worthwhile to abandon or minimize energy subsidies all over the world. Also, the ongoing transition trend in the energy sector towards green energy sources is also disrupted by reinforced production and consumption patterns through subsidies. Besides, energy subsidies grant a spurious advantage to implementing countries which impels the other countries to adopt subsidy trend to avoid arbitrage behaviour. Nevertheless, despite their detrimental effects subsidies are challenging to remove due to political reasons. As such, the G20 countries had committed to phasing out inefficient fossil fuel subsidies over the medium term in 2010 but no significant progress has been achieved thus far (Farid, 2016). Currently, similar attempts and intentions are declared by Chinese and US officials, but past experiences prove that those attempts are unlikely to achieve intended results in eliminating subsidies in the medium term. However, in recent years, the fuel prices are comparatively low compared to former years (Benes et al., 2015). Thus, it can be feasible for countries to reduce energy subsidies in this term since currently no overuse is demanded globally. Besides, the pandemic has already posed an unprecedented fiscal burden on the budget of each country and the fiscal burden of energy subsidies has been added on top of the fiscal cost of a pandemic, therefore the governments should reduce the energy subsidies in order to defer a potential fiscal downturn.

References

Acharya, R. H., & Sadath Anver, C. (2017). Implications of energy subsidy reform in India. *Energy Policy*, 453–462.

Alesina, A., Campante, F. R., & Tabellini, G. (2008). Why is fiscal policy often Procyclical? *Journal of the European Economic Association, 6*(5), 1006–1036.

Alleyne, T. (2013). *Energy subsidy reform in Sub-Saharan Africa: Experiences and lessons*. International Monetary Fund.

Andriamihaja, N., & Vecchi, G. (2007). *An evaluation of the welfare impact of higher energy prices in Madagascar*. Africa Region Working Paper Series 106 World Bank.

Bazilian, M., & Onyeji, I. (2012). Fossil Fuel Subsidy Removal and Inadequate Public Power Supply: Implications for Businesses. *Energy Policy, 45*(2012), 1–5.

Barkhordar, Z. A., Fakouriyan, S., & Sheykhha, S. (2018). The Role of Energy Subsidy Reform in Energy Efficiency Enhancement: Lessons Learnt and Future Potential for Iranian Industries. *Journal of Cleaner Production, 197*, 542–550.

Benes, K., Cheon, A., Urpelainen J., & Yang, J. (2015). "*Low oil prices: An opportunity for fuel subsidy reform*" Columbia University Center on Global Energy Policy.

Breisinger, C., Mukashov, A., Raouf, M., & Wiebelt, M. (2019). Energy Subsidy Reform for Growth and Equity in Egypt: The Approach Matters. *Energy Policy, 129*, 661–671.

Coady, D., Parry, I., Sears, L., & Shang, B. (2015). "*How large are global energy subsidies?*" IMF Working Paper Series WP/15/105.

Clements, B., Coady, D., Fabrizio, S., Gupta, S., Alleyne, T., Sdralevic, C., & Karam, P. (2013). *Energy subsidy reform: Lessons and implications*. IMF Publications.

Coxhead, I., & Grainger, C. (2018). Fossil Fuel Subsidy Reform in the Developing World: Who Wins, Who Loses, and Why? *Asian Development Review, 2018*, 35. (2).

Edgar Cooke, F. A., Hague, S., Tiberti, L., Cockburn, J., & El Lahga, A.-R. (2016). Estimating the Impact on Poverty of Ghana's Fuel Subsidy Reform and a Mitigating Response. *Journal of Development Effectiveness, 8*(1), 105–128.

Farid, M. (2016). "After Paris: fiscal, macroeconomic and financial implications of global climate change". IMF Staff Discussion Note No 16/01. http://www.imf.org/external/pubs/ft/sdn/2016/sdn1601.pdf

Granado D. J., Coady, D., & Gillingham, R. (2010). "*The unequal benefits of fuel subsidies: A review of evidence for developing countries*" IMF Working Paper Series, WP/10/202.

Inchauste, G., & Victor, D. G. (2017). *The Political Economy of Energy Subsidy Reform*. World Bank Publications.

Kyle, J. (2018). Local Corruption and Popular Support for Fuel Subsidy Reform in Indonesia. *Comparative Political Studies, 51*(11), 1472–1503.

Parry, I., Heine, D., Lis, E., & Li, S. (2014). *Getting energy prices right: From principle to practice*. IMF Washington D.C.

Rentschler, J., & Bazilian, M. (2017). Reforming Fossil Fuel Subsidies: Drivers, Barriers and the State of Progress. *Climate Policy, 17*(7), 891–914.

Robinson, L., & Mitchell, I. (2020) "*How China, Italy, the UK, and US can drive G20 fossil fuel subsidy*", Center for Global Development discussion series, https://www.cgdev.org/blog/how-china-italy-uk-and-us-can-drive-g20-fossil-fuel-subsidy-reform

Schaffitzel, F., Jakob, M., Soria, R., Vogt-Schilb, A., & Ward, H. (2020). Can government transfers make energy subsidy reform socially acceptable? A case study on ecuador. *Energy Policy, 137*.

Sovacool, B. K. (2017). Reviewing, Reforming, and Rethinking Global Energy Subsidies: Towards a Political Economy Research Agenda. *Ecological Economics, 135*, 150–163.

Töpfer, K. (2004). In A. Von Moltke, C. McKee, & T. Morgan (Eds.), *Energy subsidies: Lessons learned in assessing their impact and designing policy reforms* (1st ed.). Routledge.

Vagliasindi, M. (2013). *Implementing Energy Subsidy Reforms: Evidence from Developing Countries*. World Bank. https://doi.org/10.1596/978-0-8213-9561-5

Wattanakuljarus, A. (2021). Diverse Effects of Fossil Fuel Subsidy Reform on Industrial Competitiveness in Thailand. *Eurasian Economic Review, 11*, 489–517.

The manufacturer's authorised representative in the EU is Springer Nature Customer Service Centre GmbH, Europaplatz 3, 69115 Heidelberg, Germany. If you have any concerns regarding our products, please contact ProductSafety@springernature.com

Printed and bound by CPI Group (UK) Ltd, Croydon, CR0 4YY
30/09/2026
02225696-0010